Edward Payson Jackson

The Earth in Space

A Manual of Astronomical Geography

Edward Payson Jackson

The Earth in Space
A Manual of Astronomical Geography

ISBN/EAN: 9783337395896

Printed in Europe, USA, Canada, Australia, Japan

Cover: Foto ©berggeist007 / pixelio.de

More available books at **www.hansebooks.com**

THE EARTH IN SPACE

A Manual of

ASTRONOMICAL GEOGRAPHY

BY

EDWARD P. JACKSON, A.M.
INSTRUCTOR IN PHYSICAL SCIENCE IN THE BOSTON LATIN SCHOOL

BOSTON
D. C. HEATH & CO., PUBLISHERS
1889

PREFACE.

THIS little manual is a condensed and simplified version of a *Mathematical Geography* published several years ago. It has been prepared in compliance with the request of the late Miss Lucretia Crocker, a supervisor of the Boston Public Schools, and of many other advocates of more thorough instruction in this branch.

It is designed for Grammar Schools, and for High and Normal Schools where Astronomy is not prescribed, but where an hour a day for a few days can be spared for this most practical department of Astronomy.

The proofs have been read by Dr. J. R. Webster, Mr. C. F. King, and others. At the suggestion of several of these readers, certain passages which may be found too difficult for Grammar-School classes are marked with the character (†). The substance of these passages may be given orally by the teacher, or, if necessary, they may be omitted altogether.

Those wishing a fuller presentation of the more difficult topics are referred to the larger work mentioned above.

CONTENTS.

—◦◦•—

THE EARTH IN SPACE.

———∘o⟩⟨∘o⟨∘∘———

I. SPHERICAL FORM OF THE EARTH.

IF you were on board a steamer in the midst of the Atlantic Ocean, you would not be content with knowing all about your fellow-passengers, the internal structure of the vessel, its freight, etc. ; you would be at least equally anxious to know where you were sailing, and what countries and other objects you would pass in your course.

1. **The Earth is a Great Spherical Ship,** carrying you swiftly onward in the ocean of space. You are now asked to look abroad and see how, why, and where you are moving, and what objects you are passing at greater or less distances.

2. *What made the Earth Spherical?* — The same cause that makes the raindrop spherical, viz., *the mutual attraction of its particles.* Every particle of matter in the universe attracts, or tries to draw to itself, every other particle. This universal attraction is called the *attraction of gravitation*, or simply *gravitation.*

3. *How Attraction makes the Raindrop Spherical.* — Every one knows that drops of rain are produced

by invisible particles of cloud or vapor running together. We may imagine two or three of these particles collecting and forming a little body, which attracts more powerfully than the single particles around it.* We may then imagine the surrounding particles gathering around this body as a centre, until it becomes heavy enough to fall as a drop of rain. Now, the particles in the drop endeavor to approach as near as possible to the centre, and thus form a *sphere*, just as a party of men, in crowding around an object, form a *circle*.

4. *How Gravitation made the Earth and Other Heavenly Bodies Spherical.* — In precisely the same way. We may imagine a time when the particles of matter which they contain were scattered through space, like particles of vapor in the air, and we may imagine those particles collecting around different centres, called centres of gravitation, until spherical masses were formed of all sizes, from that of the raindrop to that of the sun itself.

How we know that the Earth is Spherical.

5. **First Proof.** — The curvature of its surface may be actually seen. When we look at a distant object upon the ocean or across a wide plain, we can see the intervening surface *rounding up* so as to conceal entirely the lower part of the object. This convexity is always found to be the same for the same distance, which could not be the case except upon a spherical body.

---

* The more matter a body contains, the more powerfully it attracts.

Whenever you are at the seaside near a great port, you may see this very satisfactorily illustrated : the more distant vessels are, the lower they seem to sink behind the convex surface of the water.

Fig. 1. The Curvature of the Earth.

A more accurate experiment consists in fixing three targets of equal height at equal distances upon a plain — as a long sea-beach — and "sighting" the elevation of the middle target above a perfectly straight line connecting the other two. The elevation is invariably found to be the same for equal distances.

6. **Second Proof.** — *Circumnavigation.* Navigators have started from a certain point and sailed constantly in the same general direction until they have finally reached the very place from which they started. Now, if the earth were of any other form than round, or if there were great edges or sudden turnings of any description in its form, these men could not have failed to discover indications of them.

7. **Third Proof.** — *The Horizon* seems both to enlarge and to sink as we ascend above the surface ; whereas, if the earth were an extended plain, our field of view would not change, whatever our elevation. The horizon is also always circular, which would not be the case if the earth's form differed very much from that of a sphere.

You may illustrate this readily as follows : Cut a small circular hole in a card, and place it upon different parts of a globe. Supposing an observer to stand in the very centre of the aperture, in each position, the circle around him represents his horizon. If some other object be taken to represent the earth, as a cube or a cylinder, it will be seen that the hole in the card must be of different forms in order to fit different parts of its surface. That part of the globe which is seen through the aperture in the card is so small as to appear perfectly flat. This explains why the surfaces of plains and of the ocean seem flat to the ordinary observer.

It may be thought that the three proofs given above do not show *positively* that the earth is spherical — that it might be of some other rounding form, like that of an egg, for example, without materially affecting the appearances described. Such a supposition is shown to be incorrect from the two proofs which follow.

8. **Fourth Proof.** — *The Weight of a Body* is very nearly the same at all parts of the earth's surface, which could not be the case if the earth were not nearly spherical, since the same body grows heavier, the nearer it approaches (on the surface) to the centre of the earth.

9. **Fifth Proof.** — *Eclipses.* In the course of the revolutions of the heavenly bodies, the earth sometimes passes exactly between the sun and the moon, casting a shadow on the latter. This shadow is *always circular*, showing that the earth is round *in every direction*. If it were round in only one direction, like a coin or a

medal, its shadow would be circular only when its flat surface exactly faced the moon. In all other positions, the shadow would vary from a straight line to an oval of different degrees of breadth, as may be easily proved by experiment.

10. **Up and Down.** — These directions are not fixed in space, like north and south, but depend entirely upon the position of the observer ; and since the earth is a sphere with observers upon all sides of it, *up* may be any direction *from* the centre of the earth, and *down*, any direction *towards* the centre. Apart from the earth or any other heavenly body, that is, in space, there are no such directions as up and down.

II. DEPARTURES FROM THE SPHERICAL FORM.

11. **The Earth's Form differs from that of a Perfect Sphere in Two Respects :** —

(1) Its surface is uneven.
(2) It is flattened at the poles.*

12. *The Amount of these Departures* is so slight that, if the whole figure of the earth could be seen at once, as by a spectator upon the moon, it would appear perfectly round, like the moon.

13. **Unevenness of the Surface.** — In order to be a perfect sphere, the earth's surface should be per-

---

* Careful measurements have recently shown that the equator also is very slightly flattened upon two opposite sides.

fectly smooth ; for hills and plains are not " equally distant from the centre."* Yet there are still higher mountains and deeper valleys on the moon, which, nevertheless, always appears to us perfectly round. All the hills, mountains, valleys, forests, cities, etc., on the earth are so extremely small, compared with its own vast bulk, that they merely serve to *roughen* its surface in a slight degree.

14. *The Principal Cause of the Unevenness.* — The earth was formerly a mass of melted matter. As the outer portions cooled, they hardened into a shell of solid rock, called the *crust.* This, in hardening, *shrunk*, and thus formed mountains and valleys, just as a smooth apple becomes *puckered* when it dries and contracts. Since then, volcanic action, which is nothing more than the *boiling* of the melted matter within the crust, has modified the form to some extent.

15. **The Flattening at the Poles** is due to the earth's rotation, which has caused those parts near the equator to bulge out, and those near the poles to sink correspondingly.

16. *The Mathematical Name* of the figure thus produced is *oblate spheroid.* The apparatus

Fig. 2. Oblate Spheroid.

shown in Fig. 3 strikingly shows the effect of rotation upon a spherical body. Two flexible hoops cross

* A *sphere* is defined as a figure all points of whose surface are equally distant from a point within called the centre.

each other at right angles, and through their points of intersection is passed a vertical rod. When the hoops are at rest they occupy the positions shown by the dotted lines ; but upon being rapidly rotated, they slide down upon the rod and assume the spheroidal form.

17. *The Amount of the Flattening at the Poles.* — The difference between the polar and equatorial diameters of the earth is about 26 miles. Each pole, then, is depressed only about 13 miles, a distance equal to only a little more than twice the height of a very high mountain.

HOW WE KNOW THAT THE EARTH IS FLATTENED AT THE POLES.

18. **First Proof.** — *Analogy.* We have, as among the proofs of its spherical form, what is called the argument of analogy ; viz., all rotating bodies are subject to the law that flattens the flexible hoops (§ 16). We know from observations of the disks of other heavenly bodies, that they are

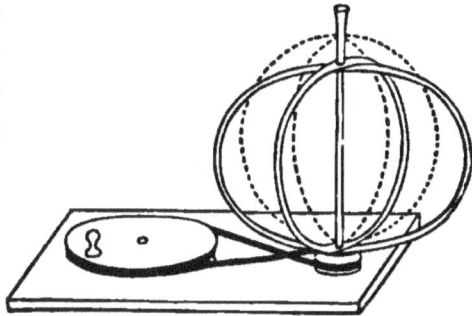

Fig. 3.   Effect of Rotation.

obedient to the law ; hence we reason that the earth must be so likewise.

19. **Second Proof.** — *Actual Measurement.* A method of measuring the curvature of the earth was

given in § 5 (Fig. 1). Another method is given in § 44. It is found that this curvature is greatest at the equator, and that it grows less and less towards the poles.

20. **Third Proof.** — *Weight.* The nearer a body above the surface of the earth is to the centre, the more it weighs. It is found that a body weighs a little more near the poles than at the equator; hence we reason that the poles must be nearer the centre.*

EXERCISES.

1. Estimating the earth's diameter at 8000 miles, what should be the thickness of a grain of sand to represent a mountain 5 miles in height, upon a globe 6 inches in diameter? *Ans.* The height of the mountain is $\frac{5}{8000}$ of the earth's diameter. The thickness of the grain of sand must, therefore, be $\frac{5}{8000}$ ($\frac{1}{1600}$) of the diameter of the globe, or $\frac{3}{800}$ of an inch.

2. What thickness should be scraped from the poles of the globe to represent the proper amount of depression? (§ 17.)

3. What should be the width and depth of a scratch upon the globe to represent a river $\frac{1}{2}$ of a mile wide and $\frac{1}{100}$ of a mile deep?

4. What would finally result if the rapidity of the earth's rotation should be indefinitely increased? *Ans.* The earth would be shattered, and its fragments would be hurled into space.

5. Of what material is the earth's axis composed?

6. Would the earth be a more comfortable or beautiful habitation for us if its form were changed to that of a perfect sphere? If the irregularities were made very much greater than they are?

---

* *Centrifugal force* would, of course, diminish the weight somewhat at the equator; but nice calculations show that this does not account for *all* the difference of weight in the two positions.

7. Suppose a cloud of dust were thrown from the earth into space with sufficient force to prevent its returning; what change would take place in the size of the mass? In its form? (§ 4.)

8. Suppose the horizon appeared to navigators sometimes circular and sometimes oval; what would such appearances imply in regard to the form of the earth? (§ 6.)

———◆◇◆———

III. LATITUDE AND LONGITUDE.

21. **Division of a Spherical Surface.** — The earth's surface contains about 200,000,000 square miles. The surface of a sphere cannot be laid out in squares like a well-planned city. How then shall we lay it out? The rotation of the earth furnishes us with certain fixed lines and points, by means of which we are enabled to lay out its surface even more easily than if it were a vast plane.

22. *Poles, Axis, Equator, etc., fixed by the Earth's Rotation.* — In a motionless sphere no point is distinguished from the rest except the point in the centre. When rotation begins, however, new relations are immediately established. Two opposite points upon its surface remain stationary, which are called the *poles*. The line connecting these points, also stationary, passes through the centre, and is called the *axis*. Points upon the surface move in circles around the axis, which are called *parallels of latitude*. These increase according to their distance from the poles, the middle and greatest parallel being called the *equator*. Circles drawn upon the earth's surface through the poles and cutting the parallels and equator at right angles, are

called *meridian circles.* Half of each meridian circle, extending from pole to pole, is called a *meridian.*

23. *Planes of the Equator, Parallels, etc.*—We may form a good idea of a plane by imagining a perfectly smooth and straight sheet of glass with indefinite length and breadth, but without appreciable thickness. Such planes we will conceive to divide the earth in different parts. The plane C, Fig. 4, cutting the sur-

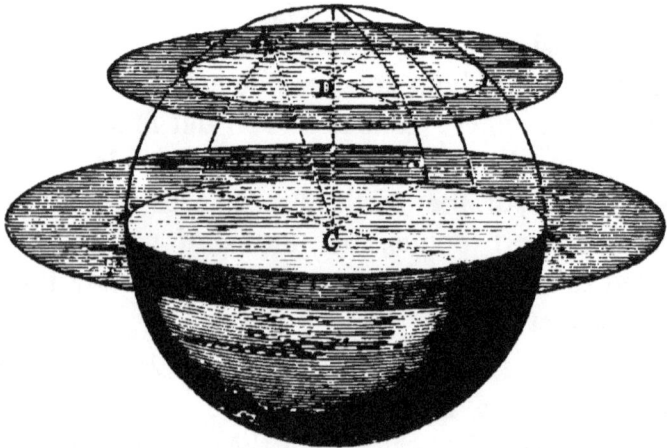

Fig. 4. **Planes of the Equator and of a Parallel.**

face through the equator, is called the *plane of the equator;* through a parallel, D, the *plane of a parallel,* etc.

The figure shows the planes extending to but a short distance beyond the earth's surface. There is, however, no limit to their extent ; the plane of the equator, for example, not only divides the earth, but it may be conceived as dividing *all space* into two equal parts. The circle in which this plane cuts the sky is called the *equinoctial* (§ 69).

24. **Latitude** is distance from the equator, measured in degrees on a meridian, either north or south. Thus, the north pole is in latitude 90° north; the tropic of Capricorn is in latitude 23½° south.

25. **Longitude** is distance from a certain fixed meridian, measured in degrees on a parallel, either east or west.

The Prime, or First, Meridian is the fixed meridian from which longitude is measured, as latitude is measured from the equator. If there were a certain meridian naturally distinguished from all the rest, as the equator is distinguished from all the other parallels, of course it would be selected as the prime meridian. But there is no such meridian; all are of the same length, and we can distinguish them only by important places through which they pass. Thus, the meridian which passes through Washington is called the meridian of Washington, and Americans sometimes measure from this as the prime meridian. More commonly, however, Americans use the English prime meridian, which passes through Greenwich. Other important nations measure from their own capitals.

26. **Having the Latitude and Longitude of a Place given,** we know its exact position upon the earth's surface, and can find it upon a map or a globe as readily as we can find a house by its street and number, or a soldier by his regiment and company.

27. *Significance of the Names.* — We can measure but 90° from the equator, while we may measure 180° from the first meridian. Hence the former distance

is called latitude (*breadth*), and the latter, longitude (*length*).

28. **The Length of Corresponding Degrees of Latitude** is the same, on whichever meridian they may be measured. Near the poles they are longer than near the equator; but the difference is so slight as to be unimportant, excepting as a proof of the spheroidal form of the earth (§ 19).

29. **The Length of Degrees of Longitude.** — Measured on the equator they are the same as degrees of latitude, viz., $\frac{1}{360}$ the circumference of the earth. Measured on any other parallel they are less, since the parallel circle itself is smaller than the equator or a meridian circle. Hence there is no fixed standard of length for degrees of longitude, which vary all the way from $69\frac{1}{6}$ miles at the equator to o at the poles.

30. **Maps and Mapped Globes.** — With the aid of this admirable system of parallels and meridians, it requires only accuracy and unwearied industry to enable geographers to represent on maps and globes the comparative position, extent, and outline, or form, of the continents, islands, countries, seas, etc., which variegate the immense surface. Should they attempt the task without the aid of these guiding lines, they would soon find themselves lost in the most hopeless confusion.

EXERCISES.

1. Can a place be farther north than the north pole? How many degrees of north latitude are there?

2. When a ship is sailing directly away from the equator — in other words, when it is "making latitude" — is it sailing

along a parallel, or a meridian? Then is latitude measured on a parallel, or a meridian?

3. Is longitude measured on a parallel, or a meridian?

4. A certain vessel was wrecked in latitude 10° south, longitude 10° west from Greenwich. Near what land was it?

5. In what longitude is New York City, measuring from Greenwich? From Washington? From Paris?

6. Is the 180th degree of longitude east, or west, longitude?

7. In what longitude are the poles?

8. If a vessel should sail directly north from the equator, steering in the same direction until it has passed over a space equal to 100°, in what latitude would it be?

9. A vessel sails due west from the meridian of Greenwich, over 200° of the parallel; in what longitude is it from Greenwich?

10. How far apart may two points be, and yet be in the same latitude?

11. How far apart may two points be, and yet be in the same longitude?

12. How many meridians may be drawn through a parallel?

13. How many parallels may be drawn through a meridian?

14. How many miles in a degree of latitude?

15. What is the length, in miles, of a degree of longitude measured on the equator?

16. About what part of a mile does a degree of longitude measure on the parallel seven miles distant from the pole? (*Consider the space within the circle flat, and multiply the diameter of the circle by* 3½, *to obtain its approximate circumference.*)

IV ZONES.

31. **Meaning of "Zone."** — If the space between two parallel circles upon the surface of a sphere be distinguished from the rest of the surface, it will pre-

sent the appearance of a *belt* encircling the sphere; hence the name *zone* (belt).

32. Number and Names of the Zones. — The surface of the earth is divided, by four parallel circles, into five zones, viz., *North Frigid, North Temperate, Torrid, South Temperate,* and *South Frigid.* The four dividing circles are the *Tropics* and the *Polar Circles.*

THE ZONES ARE NATURAL DIVISIONS OF THE EARTH'S SURFACE.

33. The Torrid Zone. — The sun's vertical rays do not always fall upon the equator. Sometimes they fall upon the parallel $23\frac{1}{2}°$ north of the equator — the Tropic of Cancer; at other times, upon the parallel $23\frac{1}{2}°$ south of the equator — the Tropic of Capricorn; and, in the meantime, upon every circle between these two. But they never fall farther north or south than the tropics. (These northward and southward movements of the sun will be described more fully, and their causes explained, in a future section.) Now, the solar light and heat are the most intense where the vertical rays fall; hence these rays mark off a zone of the earth's surface 47° in breadth, divided from the rest of the surface by the tropics. This is called the *Torrid* (or *hot*) *Zone.*

34. The Frigid Zones. — For the same reason that the vertical, or hottest, rays of the sun do not always fall upon the equator, the most oblique, or *coldest*, rays do not always fall at. the poles. And as the vertical rays range $23\frac{1}{2}°$ from the equator, so the most oblique

rays range 23½° from the poles. Hence these rays
mark off two sections of the earth's surface, each 47°
in diameter, and divided from the rest by the Arctic
and Antarctic circles. These are called the *Frigid*
(or *cold*) *Zones.*

35. **The Temperate Zones.** — Between the Torrid
and the Frigid lie the *Temperate* Zones, upon which
neither the vertical nor the most oblique rays of the
sun fall, that is, neither the hottest nor the coldest
rays, hence their name.

Exercise. — What is the breadth of the Temperate Zones?

36. **The Difference between the Climate of the
Torrid and that of the Frigid Zones** is so great that,
if animals belonging to one should be carried to the
other, they would soon perish ; and, while the rankest
luxuriance of vegetation grows in one, there are but a
few hardy shrubs and mosses to vary the eternal snows
of the other.

37. *The Reasons for this Great Difference* may be
easily understood from Figs. 10 and 11, and the para-
graphs relating to them.

38. *Gradual Decrease in Temperature from the
Equator to the Poles.* — Representations of the earth,
with the zones painted upon its surface, imply very
abrupt changes at the dividing circles ; but, of course,
no more sudden change would actually be experienced in crossing these dividing circles than in crossing any other parallels.

V. DIMENSIONS AND DISTANCES.

39. **The Circumference of the Earth** is, in round numbers, 25,000 miles.

40. **The Diameter** is a little less than one-third of the circumference, or about 8,000 miles. (See Appendix, I.)

41. **The Earth's Distance from the Sun** varies slightly during the year, the greatest distance (*Aphelion*, § 75) being estimated at 94,500,000 miles ; and the least (*Perihelion*), at 91,500,000 miles. The average or mean distance is, therefore, 93,000,000 miles. (See Appendix, II.)

42. **The Earth in Space.** — The earth seems to us an immense body, yet it is one of the smallest of the bodies that roll in space. In fact, the whole solar system taken together is but a mere point when compared with the universe.

As the student of botany regards some insignificant plant as only an individual of an innumerable species, so let us regard the earth as only a small specimen of countless myriads formed on the same general plan.

How we know the Magnitudes and Distances of the Heavenly Bodies.

43. **Measurement of the Earth.** — *First Method.* The amount of curvature of a small portion of the surface may be measured as shown in § 5, Fig. 1, and from this arc the whole circumference may be found.

Exercise. — A peak of the Andes 4 miles high is just visible on the Pacific Ocean at the distance of 178⅓ miles. This gives an arc of a circle equal to 2.58 degrees. How many miles in 360 degrees? *Ans.* 24,872 + miles.

44. *Second Method.* — The north star remains apparently motionless in the sky, excepting when we move towards or from it. If we sail towards it over one degree of latitude, it seems to ascend one degree towards the zenith. To cause this appearance, it is found that a ship must sail about 69⅓ miles directly north — one degree of latitude. Multiplying this by 360, we have the circumference, 24,900 miles.

45. †[1] **Estimation of Distances.** — In giving us *two eyes*, nature has enabled us to measure very accurately the distances of objects immediately around us.

Hold your finger very near your eyes (Fig. 5). To look directly at it, you must turn your eyes inward, in other words, look " cross-eyed." Now look beyond your finger at a wall or a window. The finger will appear in two different positions against the wall or window. Close the right eye, and it will appear against A. Close the left eye, it will appear against B. The farther you remove your finger from your eyes, the less you have to turn your eyes inward to look at it, and the shorter grows the line AB on the window (called the *parallax*). In general, the less we have to turn our eyes inward, or towards each other, to look at an object, the farther off we judge it to be ; in other words, the less its parallax becomes.

Exercise. — How far off must a point be in order that the eyes must look *in exactly parallel lines* to look directly at it?

---

[1] See Preface.

The farther apart our eyes are, the more, of course, we have to turn them inward to look at an object, and the better judges of distance we are. Our eyes are too near together to serve in this way for great distances. Thus, the heavenly bodies all appear at the same distance from us, although their real distances are immensely different. Still, we can apply the principle in a different way. Fig. 6 represents two men

Fig. 5. **Estimating Distances.**

measuring the distance of a balloon. Knowing their own distance apart (AB, the " base line "), and finding how much they have to turn towards each other to look directly at the balloon, they can, by a mathematical process too difficult to explain here, calculate its distance precisely as you judge of the distance of your finger by the angle at which you have to incline your eyes inward to see it. Observe that the arc CD (Fig. 6) in the sky is the *parallax* of the balloon, just as AB (Fig. 5) is that of the finger on the window.

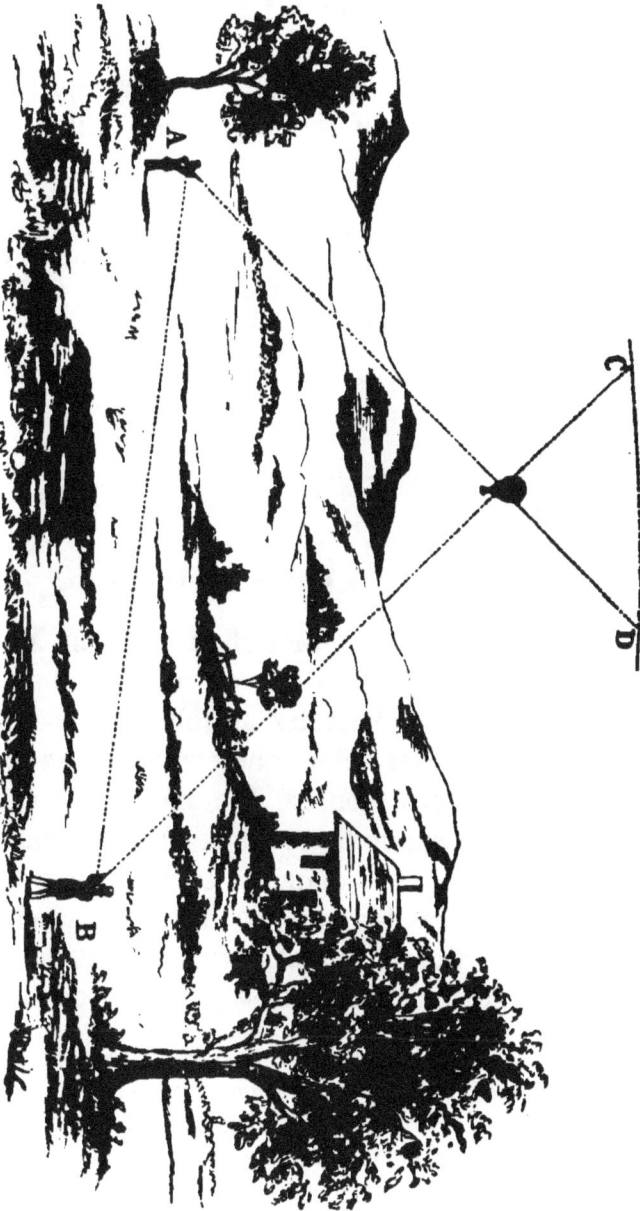

Fig. 6. Measuring the Distance of a Distant Object.

The farther away an object is, the greater must be the base line (AB, Fig. 6) to measure its distance. In the cases of the sun and the nearer planets the base line used is half the diameter (radius) of the earth — 4,000 miles.

Exercises. — 1. If ABC (Fig. 7) represents the earth, and S the sun, what line represents the sun's parallax in the sky?

2. If the sun were removed farther from the earth, would this line increase or diminish?

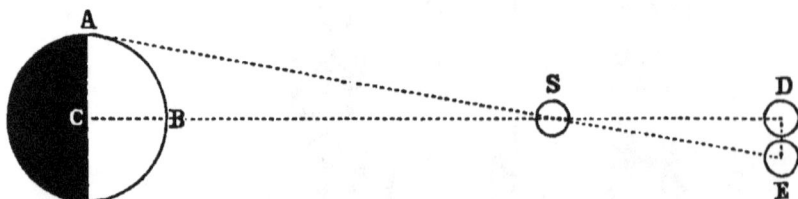

Fig. 7. The Parallax of the Sun.

46. †[1] **Measuring the Size of the Sun and Moon.** — Having thus found the distances of these bodies, we can measure their diameter as follows : Place a circular piece of paper before one eye at such a distance that it will exactly conceal the moon, for example, from view. Now a circle twice as far away would have to be of just twice the diameter to fill the same space in the eye ; one as far away as the moon would have to be as many times as great in diameter as the moon's distance is greater than that of the paper circle.

Exercise. — If a paper disk 1 inch in diameter be pasted on a window-pane in view of the full moon, a spectator will have to withdraw $9\frac{7}{27}$ feet from the paper disk in order that it may just conceal the moon from his sight. Required, the moon's diameter.

---

[1] See Preface.

Ans. The moon's distance being 240,000 miles, or 136,857,-600 times greater than that of the paper disk, its diameter must also be the same number of times greater than that of the paper disk, — 136,857,600 × 1 inch = 2,160 miles.

———◆◇◆———

VI. THE SUN'S RAYS AND THE EARTH'S ATMOSPHERE.

47. **Light is the Means of Sight, although itself Invisible.** — We often speak of "seeing light"; but it is not light that we see, but the various *objects* which send light to our eyes. Light is not a substance; it is only a *means* which, by affecting our eyes, enables us to see substances. If there were no substance in view, we should be surrounded by darkness, even though the space around us were filled with rays of light.

48. *Examples showing that we do not see Light.* — 1. If we stand before a brilliantly lighted window upon a dark evening, the rays from within will pour upon us in a flood, and the space between us and the window will seem bright with the light. But if we move to a corner of the building so that the window itself shall not be within the range of vision, the space in front of it, if the air is perfectly clear, will seem as dark as if the window were unlighted; we cannot see the *rays* pouring out into that space. If, however, some *substance* move into the space, as a passing carriage or even a cloud of mist, it will be suddenly lighted up. The *rays* will enable us to see the *substance*.

2. At midnight, when the sun is far below the horizon, his rays must, of course, shoot up on all sides of the earth, as shown in Fig. 8 ; and if these rays could be seen, they would present the appearance of a dazzling shower pouring up from the horizon on all sides, causing the night to be nearly as bright as the day. At midnight, however, we see no evidence of the sun's rays, unless there be some *substance* above us, like the moon or the planets, to receive the rays and throw them back to us.

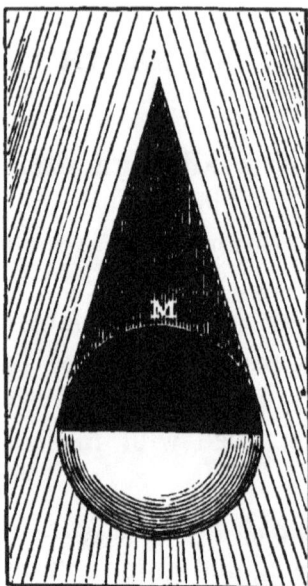

Fig. 8.

49. **Air* in Large Quantities is Visible,** and is of a pale blue color. We look upward in the daytime, and see what seems to us an immense flood of pure light. Of what is this vast illuminated ocean composed, whose upper surface seems a shell of pale blue?

It cannot be *light* that we see, for we have shown light to be invisible. Plainly then it must be some *substance* lighted by the sun's rays, just as a cloud of mist would be lighted by a candle.

That substance is the *air*. If it were removed, the flood of light would disappear, and we should see nothing above us, even at noon, but a black, measure-

---

* Including all the various solid and liquid particles which the atmosphere contains.

less abyss, with the sun glaring in the midst, and the stars and the moon as plainly visible as they are now at night.

Fig. 9. The Earth's Atmosphere illuminated by the Sun's Rays.

50. *Why we do not realize that the Air is Visible.* — This is because we have nothing more transparent than itself with which to contrast it. If a great ball of air

could be suspended in empty space beyond our atmosphere, we should see it shining at night like a
planet.

Fig. 9 shows the air thus lighted by the sun's rays.
The *halo*, represented as half surrounding the earth,
may be regarded as a picture of the flood of light
which we see above us in the daytime, and which
would disappear if the air were removed.   If we could
ascend above this, we should see the black, starry
space beyond, as is proved by those who make balloon
ascensions or climb lofty mountains.   These men
describe the sky growing darker and darker, as they
leave more and more of the atmosphere below them,
until the stars become visible even at noonday.

Fig. 8 is an example of the manner in which light is
usually represented.   It will be understood, however,
that only the *directions* of the rays are shown by the
straight lines, as the directions of the equator, ecliptic,
etc., are shown by curved lines.

51. **The Speed of Light** is about 186,000 miles
a second.   It requires, therefore, a little more than
eight minutes for a ray of light to reach us from the
sun.

GRADUAL CHANGES IN LIGHT AND HEAT DURING THE
DAY AND THE YEAR.

52. **Twilight** is the gentle sunlight that plays around
us before sunrise and after sunset.   It is nothing more
than the gray border of the " flood of light," described
in § 49 and represented in Fig. 9.

How Twilight is produced. — Long before the sun's direct rays reach us they shine upon the upper regions of the air in the east, and produce the first gray streaks of dawn.* We see this brightened air in the distance just as we see the tops of distant mountains lighted up before sunrise and after sunset. As the sun rises higher and higher towards the horizon, more and more of the air above our heads becomes lighted up, until at the appointed instant his broad disk bursts into view.

53. **The Day.** — Even then we have not the full light of day. We may gaze upon the very face of the sun, and scarcely feel his warmth. As he makes his sublime ascent he becomes more and more powerful until he reaches his culmination, after which his power as gradually diminishes, till the second twilight has faded away into the darkness of night.

54. **The Year** presents similar gradual changes in light and heat, and the changes of both the day and the year are due to the same cause, viz., differences in the direction of the sun's rays.

TWO REASONS WHY DIFFERENCES IN THE DIRECTION OF THE SUN'S RAYS MAKE DIFFERENCES IN THEIR POWER.

55. **First:** The *more slanting* the rays, the *greater the surface* over which they are scattered, and hence the *less intense their power.*

Fig. 10 represents three sheaves, or bundles, of the sun's rays, striking the earth at three different periods

---

* Twilight begins when the sun is about 18° below the horizon.

of the day. At noon the rays are nearly vertical, and fall
upon the surface between C and D. In the middle of
the afternoon they are inclined, and are spread over a
greater surface, DE. At sunset the lower side of the

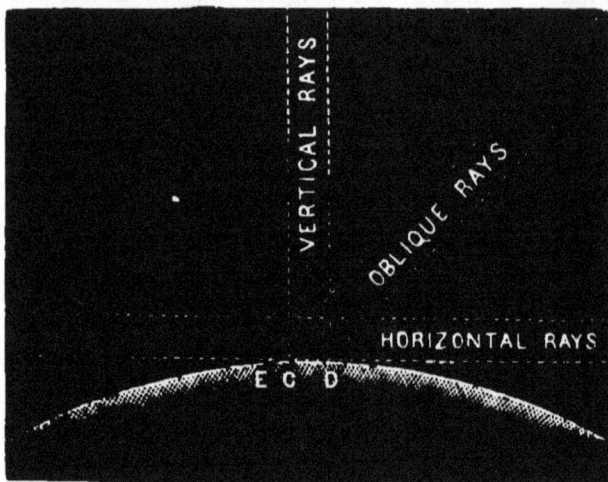

Fig. 10. Dispersion of Rays over Surface.

sheaf just touches the surface, while most of the rays
themselves are lost in the space beyond.

56. **Second:** The *more slanting* the rays, the *more
air* they must pass through, and therefore *the more
they are interrupted and absorbed.*

We have already learned that the air is not perfectly
transparent; it interrupts the light as glass interrupts
it, though not in so great a degree. A clear pane of
glass seems to admit as much light as an open space
of equal extent; but if it were a hundred times as
thick it would admit scarcely any light. So if the air
were sufficiently increased in quantity we should be
left in utter darkness, as if we were at the bottom of

the ocean. When we look at the sun in the horizon, we see him through an immense mass of air ; hence he sometimes appears of a dull red color, as if seen through smoked glass.

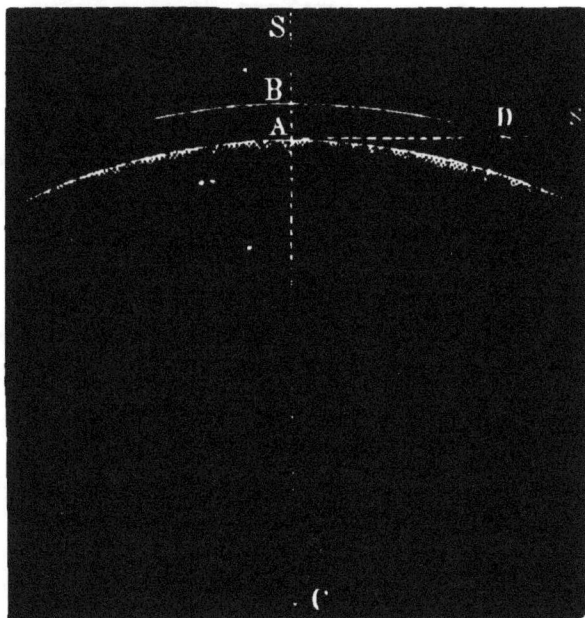

Fig. 11. Interruption of Rays by the Atmosphere.

57. In Fig. 11, SBA represents a noon ray of sunlight, and S′ DA one at sunrise or sunset. If the curved line BD represent the outside of the atmosphere, fifty* miles above the surface of the earth, it is

---

* Fifty miles is the height formerly assigned to the atmosphere, although it probably stretches off, in a state of extreme rarity, many hundred miles beyond this limit, gradually shading off into nothingness. Its *great mass*, however, is within eight or ten miles of the earth's surface, the more elevated portions comparing with the lower very much as the vapor of the ocean compares with the ocean itself.

evident that the rays must penetrate through very much more air when the sun is in the horizon than when he is overhead.

———◦◊◦———

VII. THE EARTH'S DAILY MOTION.

58. **The Earth performs Two Motions.** — It rotates upon its axis once in twenty-four hours, and it revolves around the sun once a year.

59. *The Motions are Permanent.* — Nothing made by man will continue in motion indefinitely, unless new force is repeatedly applied to it. A watch will stop unless it is wound at regular intervals, and even then it will finally wear out. But the earth never wears, and will never stop unless some great change takes place in nature to produce such a result.

60. *What keeps the Earth moving?* — A top will spin much longer on a smooth surface than on a rough one ; it will spin scarcely a second of time under water. Having been set in motion, the length of time during which it will continue moving depends, *first,* on the amount of friction between the peg of the top and the surface on which it spins ; *secondly,* on the density of the medium in which it spins. Suppose the top to be spinning in a perfectly empty space without friction or any other resistance to overcome, how long will it continue in motion? — a day? a year? What will then cause it to stop? It will require as much force to destroy its motion as was required in

the first place to produce it, and unless that force is applied, it will continue spinning forever.

The earth is like a huge top in precisely similar circumstances. It rotates in empty space, and there is no friction between its surface and any external surface, as there is between the peg of the top and the floor. But does not the air resist the motion of the earth as it resists that of the top? By no means; the air is a *part* of the earth — a thin covering — and, like the ocean, is carried around with it.

The same principle applies to the earth's motion around the sun as to its rotation. It continues undiminished simply because there is nothing to *resist* it.

61. The Effects of the Earth's Rotation : —

(1) Alternation of day and night (§ 62).

(2) Determination of an axis, equator, etc. (§ 22).

(3) Flattening at the poles (§ 16).

(4) Apparent rotation of the heavens in the opposite direction (§ 66).

62. The Alternation of Day and Night. — As the sun shines on only one-half the earth's surface at a time, the other half must be in darkness. If there were no motion one half would be in constant day and the other half in constant night; but the rotation of the earth brings each half, in turn, into the light and shade.

EXERCISES.

1. When it is noon at Boston, where on the same parallel is it sunrise? Sunset? Midnight? [In the questions 1 and 2 the sun is supposed to be over the equator.]

2. When it is sunrise at London, where on the same parallel is it noon? Sunset? Midnight?

3. Over how much of a meridian is it noon at the same instant?

4. Over how much of a parallel is it noon at the same instant?

5. How many miles an hour does a body at the equator move around the earth's axis?

6. If a man or an animal should perform this rapid motion *through* the air would he be likely to feel the resistance? *Ans.* The effect would be very much greater than that of the most violent hurricane.

7. Then does the earth move through the air, or does the air move with the earth?

8. What is the rate of motion at the poles?

9. Is the rate of motion on the parallel ten miles from the poles very rapid, or slow?

63. **Belief of the Ancients.** — The ancients generally supposed that the earth is perfectly motionless, and that the sun, moon, and stars perform daily revolutions around it.

Insensible Motion often causes a Similar Delusion. — The motion of a balloon through the air is so extremely gentle, however rapid it may be, that if one closes his eyes or looks only at the sky it seems motionless ; and upon looking downward the sensation is strong that the earth is *falling away* from the balloon, rather than that the balloon is rising above the earth. Although we are totally insensible of the earth's motion, yet we feel that it would be almost as absurd for us to regard the earth as stationary and the heavens in motion around it, as for the aëronaut to regard his balloon as fixed and the earth descending below it. But we are not obliged to content ourselves with mere probabilities ; the earth's rotation is proved to a *positive certainty.*

How we know that the Earth rotates.

64. First : We have seen on what principle the distances of the heavenly bodies may be measured. We know that the stars are so distant that their light, whose motion is inconceivably swift, requires *years* to reach us ; yet they *seem* to move around the earth once in twenty-four hours. We know that it is impossible for them actually to perform such motions. The only way, therefore, by which the appearances can be produced is by the earth itself turning on its axis.

65. Secondly : When a grindstone is rotating rapidly, it will throw drops of water in the direction in which it is rotating ; if, for example, its upper surface is moving toward the east, it will throw the drops eastward. *The earth does precisely the same thing.* A stone dropped from the top of a high cliff always falls a little *east* of a verti-

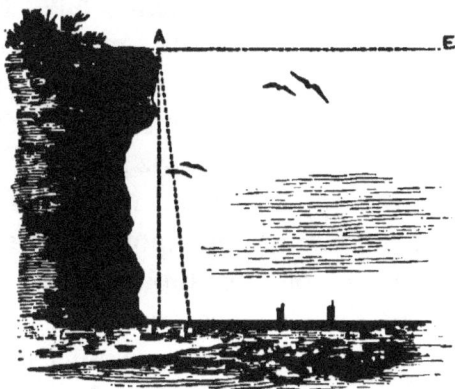

Fig. 12.   Proof of Earth's Rotation.

cal line ; that is, it is *thrown a little eastward* by the earth's rotation (Fig. 12).

If the earth's rotation were as rapid in proportion as that of the grindstone which throws off drops of water, objects on its surface would be thrown off in the same way. A stone, for example, would fly from the cliff (Fig. 12) in the horizontal line AE, towards the east. As the rotation is not nearly so rapid, however, the stone's departure from the vertical line AC in falling is comparatively slight. Thus the distance CD furnishes mathematicians with an independent means of determining the rate of the earth's rotation.

(For Foucault's famous experiment proving the earth's rotation, see Appendix III.)

APPARENT DAILY MOTION OF THE HEAVENS.

66. Different Rates of Apparent Motion of the Stars, etc. — Among the effects of the earth's daily motion is the apparent daily rotation of the starry sphere (with the sun, moon, etc.) in the opposite direction, from east to west. The same principles apply to this apparent motion that apply to the rotation of any sphere, viz., the poles (the north star* and the opposite point in the heavens) remain stationary, merely turning upon themselves as upon pivots, while the rate of apparent motion increases according to the distance from the poles, being swiftest at the *celestial equator*, or *equinoctial*, which lies over the earth's equator as the celestial poles stand over the earth's poles. Stars near the pole star seem to move in small circles around it every twenty-four hours, precisely as icebergs are carried by the earth's rotation

---

* The north star is at a very slight distance ($1\frac{1}{2}°$) from the true north pole of the heavens. Consequently, it seems to describe a minute daily circle around it.

around the north pole, only in the opposite direction ;
stars farther off describe larger circles, those over the
equator describing the largest of all.

67. **How to observe the Apparent Rotation of the
Heavens.** — Most persons know where to look among

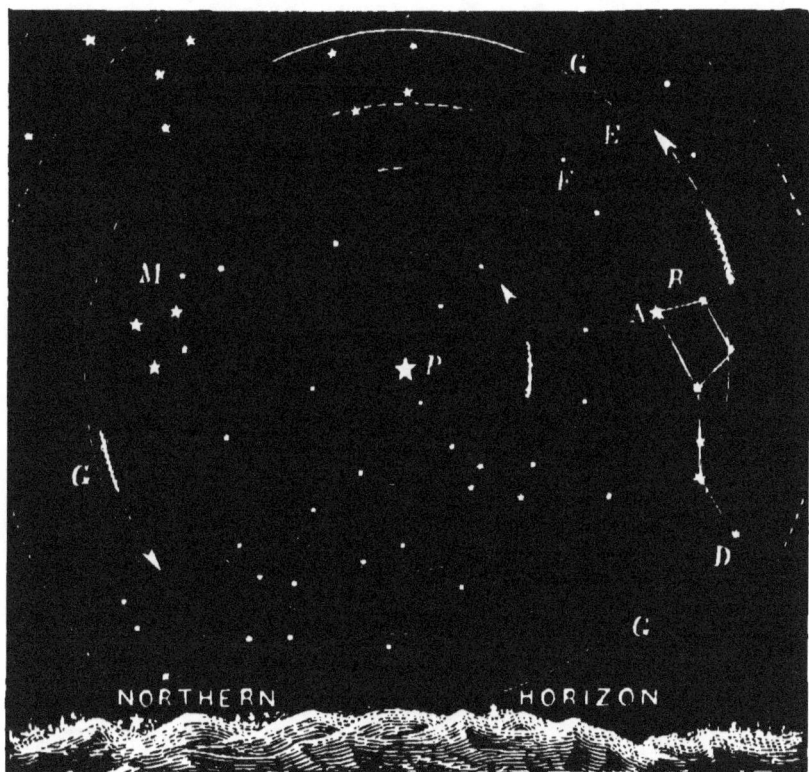

Fig. 13. Apparent Daily Motion of the Heavens.

the stars for the "*Dipper.*" It is represented in Fig.
13, ABD. The two stars A and B are called "pointers," because they seem to *point* to the pole star P.

The circles in the figure show the apparent revolution of the stars, or the rotation of the celestial sphere in the direction of the arrows, from east to west.

Now, on some clear evening carefully observe the relative positions of the Dipper and some other cluster, as *Cassiopeia*, M, which resembles an irregular W in form. The book can easily be held so that the stars in the figure shall correspond in position with those in the sky. Observe also the positions of certain bright stars overhead and near the eastern and western horizon. Compare their positions again about three hours afterwards. The north star will not appear to have moved, the Dipper will have moved through $\frac{3}{24}$, or $\frac{1}{8}$, of the whole circle, and the pointers will be at E and F. All the other stars will also have moved through $\frac{1}{8}$ of their circles, those that were in the eastern horizon having ascended, those that were overhead having descended toward the west, and those that were in the western horizon having set.

68. *Circles of Perpetual Apparition and Occultation.* — The stars within the heavily marked circle G, Fig. 13, never pass below the horizon by day or by night. This circle is, therefore, called the circle of *Perpetual Apparition* (appearance). What represents the circle of perpetual apparition in Fig. 17? The south pole of the heavens is as far below the horizon as the north pole is above it ; there must, therefore, be a circle around the south pole corresponding to G in which to us the stars never rise. This is called the circle of *Perpetual Occultation* (concealment). Both these circles increase as we move towards the poles, and diminish as we move from them.

69. Daily Motion of the Heavens as seen from Different Points of the Earth's Surface. — Fig. 14 represents the earth in the centre of the hollow sphere of the heavens; PP' are the poles of the heavens

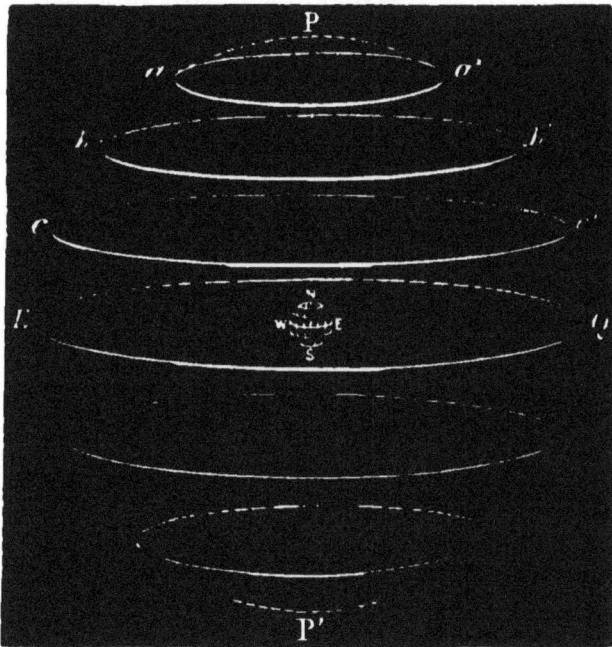

Fig. 14. Earth in Centre of Celestial Sphere.

directly over the poles of the earth, and EQ is the celestial equator, or equinoctial, over the earth's equator.

From the North Pole. — If we should stand at N, Fig. 14 or Fig. 15, our horizon would coincide with the equinoctial; the north star, P, would be directly overhead; and the other stars, with the sun, moon, and planets, would seem to move in circles around us — *aa'*, *bb'*, *cc'*, etc.

From the Equator. — If we should stand at E, Fig.
14 (hold the page so that Q shall be uppermost), or
at X, Fig. 16, the north star would be exactly in the
northern horizon, and the other stars, sun, moon, etc.,
would seem to move in circles perpendicular to the
horizon.

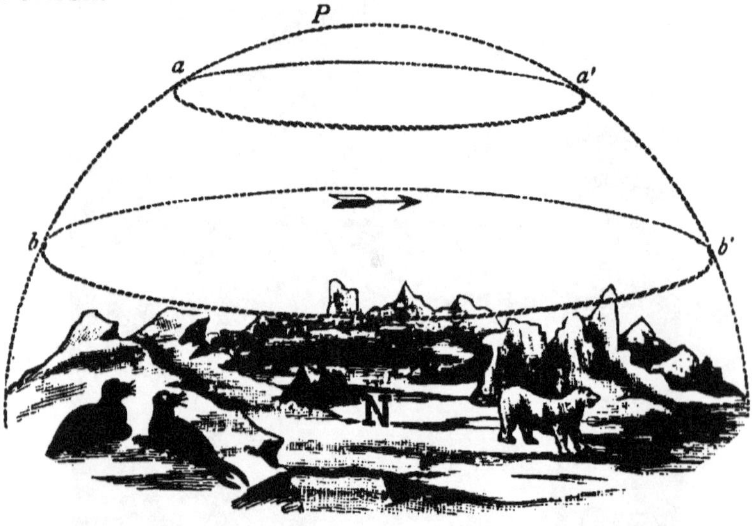

Fig. 15.   Daily Motion of the Heavens as seen at the North Pole.

From a Point between the Equator and a Pole. — It
we should stand at a point between the equator and
the north pole, as at X, Fig. 17, the north star, P,
would be seen at a distance above the horizon corresponding to our distance from the equator (§ 71, see
also Fig. 13) ; and the other stars, sun, moon, etc.,
would seem to move in circles oblique to the horizon.

70. **How Long does a Heavenly Body remain
above the Horizon?** — *Questions :*

    1. During what portion of the twenty-four hours of

the day does the sun, moon, or a star, if above the horizon at all, remain above, at the north or the south pole? (See Fig. 15.)

2. At the equator? (Fig. 16.)

3. During what portion of the twenty-four hours does a star at *o*, Fig. 17, remain above the horizon?

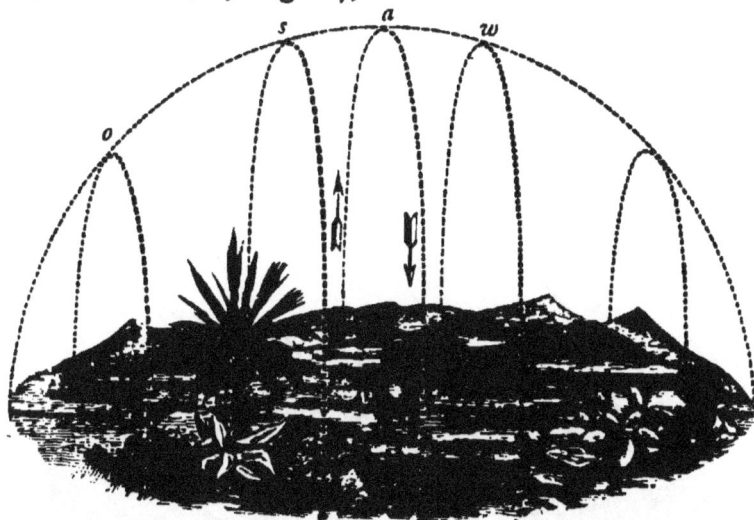

Fig. 16. Daily Motion of Heavens as seen at the Equator.

a star at *a*, over the equator? at *s*, north of the equator? at *w*, south of the equator?

4. When the sun is above the horizon at the north pole, does he set during the twenty-four hours? (*b*, Fig. 15.)

5. Is the sun ever more or less than twelve hours above the horizon at the equator, whether he be directly over the equator, or north or south of it? (*s, a, w*, Fig. 16.)

6. How does day compare with night in length

when the sun is at *a*, Fig. 17? at *s*? at *w*? (Compare with Fig. 24.)

7. Suppose the sun ever appeared as far north as *o*, Fig. 17, what portion of the twenty-four hours would be day and night respectively?

71. **The Distance of the North Star above the Horizon at any Place, measures its North Latitude.**

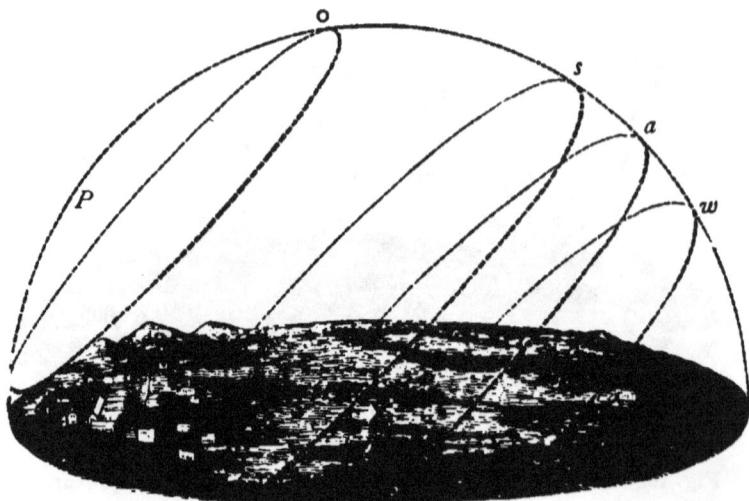

Fig. 17. Daily Motion of Heavens as seen at Latitude about 40°.

— If we are at the equator, we see the north star *in* the horizon, showing that our latitude is o. If we are at the north pole, we see the north star in the zenith, or 90° above the horizon, showing that our latitude is 90°. If we are at latitude 40° north (X, Fig. 17) we see the north star P 40° above the horizon, etc.

EXERCISES.

1. What represents the circle of perpetual apparition in Fig. 15? Fig. 16? Fig. 17?

2. If the earth turned on its axis from east to west, in what direction would the heavens seem to move?

3. Would there be anything of importance to distinguish the north star from the rest of the stars if the earth did not rotate? Would *any* of the stars appear to move?

4. How far south must we go in order to see the south pole of the heavens?

5. The crew of a ship see the north star above the horizon, one-fifth of the distance from the horizon to the zenith; in what latitude are they? (§ 71.)

6. At a time when the sun is known to be over the equator, a ship's crew see it, at noon, 10° south of the zenith; in what latitude are they?

7. Then, may latitude be determined from the sun, as well as from the north star?

———•◦•———

VIII. THE EARTH'S YEARLY MOTION.

72. What makes the Earth move around the Sun? — If a stone be attached to an elastic cord and swung around the hand, not only will the earth's revolution around the sun be illustrated, but also the *two forces which produce the revolution.* The force exerted by the hand tends to throw the stone in a direction *from* the hand ; and if that force should cease the elastic cord would pull the stone *towards* the hand. The cord prevents the stone from moving from the hand, and if the cord should break the stone would fly off in a straight line in the direction in which it happened to be moving when the cord parted.

If, for example, the cord should part when the stone reached A, Fig. 18, the latter would fly off in the line AB ; at C it would take the direction CD, etc.*   But so long as the cord remains unbroken — since the

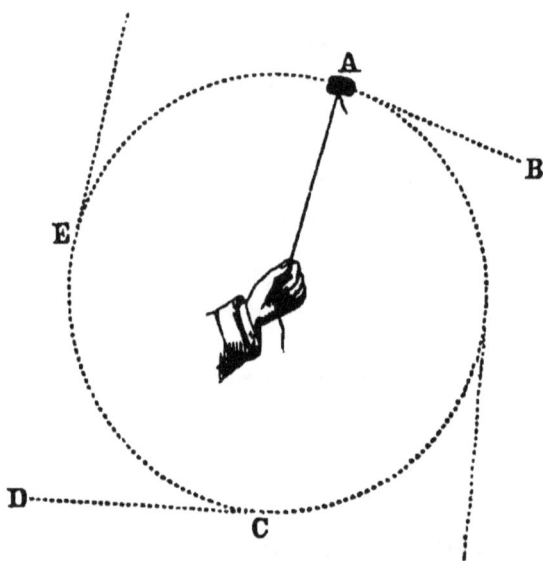

Fig. 18. Centripetal and Centrifugal Forces.

stone cannot move either towards or from the hand — it must take a direction *between*, or in a curve around the hand.

In like manner, the force which was first communicated to the earth tends to cause it to move onward in a straight line, while the sun's attraction tends to draw it to the sun ; but the two opposite forces are so

* The lines AB and CD are *tangents* to the circle ACE.

adjusted that it moves in a nearly circular pathway around the sun, which is called its *orbit.*

73. *Centrifugal and Centripetal Forces.*—The force which tends to move a body in a straight line is called, for want of a better term, *centrifugal;* that which tends to draw it from this straight line into a curve is called *centripetal.*

EXERCISES.

1. Which of these forces prevents the stone from being drawn to the hand?
2. Which prevents the stone from flying from the hand?
3. Which prevents the earth from falling to the sun?
4. Which prevents the earth from flying off into space?

74. **Form of the Earth's Orbit.**—If the two forces, centrifugal and centripetal, were exactly balanced, and no disturbances were made by the attraction of other planets, the earth's orbit would be a perfect circle; but such is not the case; its real form is that of an oval, or *ellipse,* with the sun in one focus.

75. *Perihelion and Aphelion.* — The earth's orbit being an ellipse, our distance from the sun varies slightly at different times. When the earth is nearest the sun it is said to be in *perihelion;* when most distant, in *aphelion.* (§ 41.)

76. **Both Daily and Yearly Motions are in the Same Direction;** viz., from west to east. The relation of the two motions may be fixed in the memory by associating them with those of a rolling ball or a carriage wheel, which may be said to rotate in the *same. direction* in which it advances.

77. **Principal Effects of the Earth's Revolution around the Sun.** — 1. The apparent yearly revolution of the sun around the earth, through the twelve Signs of the Zodiac.

2. The Change of Seasons. This is a combined result of the yearly motion and the inclination and unchanging direction of the earth's axis.

HOW WE KNOW THAT THE EARTH REVOLVES AROUND THE SUN.

78. **Cause of the Sun's Apparent Yearly Motion.** — As the earth's real motion upon its axis causes the whole heavens to seem to move around the earth once a day, so the earth's real motion around the sun causes the sun to seem to move around the earth once a year.

79. † *In what the Sun's Apparent Yearly Motion consists.* — The apparent daily revolution of the sun around the earth has nothing to do with its apparent yearly revolution. The former is an apparent movement of the sun *in company with the whole heavens,* while the latter consists in its *apparent changes of position among the stars.*

The stars on account of their immense distances seem as immovably fixed in the immense hollow sphere that surrounds us as if they were silver nails in a blue ceiling ; and they appear to turn with that hollow sphere once a day, as the nails would move with the ceiling if the latter should perform a rotation.

• But the sun, although in reality a star, does not appear thus fixed on account of its comparative nearness to us. As you ride swiftly along in a railway car, you notice that objects in the landscape appear to move swiftly or slowly past you, according to their comparative nearness or distance ; distant mountains appear motionless, like the fixed stars. So, if we could see the sun and stars together in the daytime we should see it slowly creeping past them from west to east, as we see from our car window a tree or a house creeping past the distant mountains. If it appears beside the star *a*, Fig. 21, at sunset to-day, it will appear at *b*, a little east of (above) that star, at sunset to-morrow, when the star *a* will be just below the horizon. The next day at sunset the sun will appear still farther east, so that the point *b* will then be below the horizon, and so on, until, in 365¼ days, the sun will appear to have made an entire circuit around the heavens, and to have returned to its starting point beside *a*.

80. † *We may observe the Sun's Apparent Motion among the Stars as accurately as if they were visible in the Daytime.* — 1. If we observe what stars are just above the western horizon as soon after sunset as they become visible (the uppermost stars in Fig. 21 for example), we shall find at the same time to-morrow that these stars have descended somewhat, which will show that the sun has approached them ; that is, that he has moved a little *eastward*.

2. Since the sun is on the meridian upon the other side of the earth at midnight, certain stars on our meridian must be *exactly opposite* the sun ; and, as *different* stars appear on our meridian at midnight

night after night, the sun must be opposite different
stars at different times.

For example, when the earth is at *a*, Fig. 19, we see
the stars at C on the meridian at midnight, and we
know that the sun is on the other side of the earth,

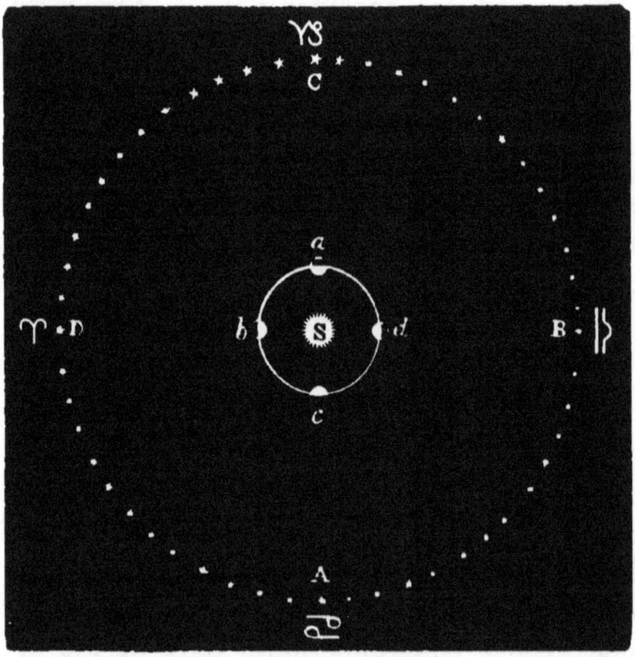

Fig. 19. The Earth's Real, and the Sun's Apparent
Yearly Revolution.

opposite these stars, appearing to observers upon the
opposite side of the earth in that part of the heavens
denoted by A.

Three months afterward, when the earth has moved
to *b*, we shall see the stars at D on the meridian at
midnight, and shall know that the sun is opposite
them.

EXERCISES.

1. To what point will the earth have moved three months later?

2. What stars shall we then see on the meridian at midnight?

3. Among what stars would the sun seem to be at this time, if we could see the stars by daylight?

4. As the earth moves from *c.* to *d,* the sun seems at the same time to move past what stars?

5. Would the sun seem to make this movement if the earth remained motionless?

6. If the sun seems to move entirely around the heavens in a year, through what part of the circle does he seem to move in a day? — *Ans.* Nearly one degree.

81. Other Heavenly Bodies move among the Stars. — The moon and planets also change their positions among the stars from day to day, moving generally in the same direction in which the sun moves ; viz., from west to east. But here a most important distinction must be made ; the moon and planets *really* move, while the sun's motion is only *apparent.*

82. The Zodiac is that great zone, or belt, of the heavens within which the sun, moon, and planets are seen to move. It is sixteen degrees in width, eight degrees each side of the ecliptic (§ 83). The stars within the zodiac are divided into twelve constellations, from which the twelve Signs of the Zodiac are named. (Appendix IV.)

83. The Ecliptic is the earth's real yearly path, or sun's apparent yearly path, through the heavens. It lies along the middle of the zodiac, as seen in Figs. 20 and 21.

† Fig. 21 represents the sun setting, with the stars visible, as they would be if it were not for the atmosphere (§ 49).   The space within the two oblique lines on either side of the sun represents a portion of the

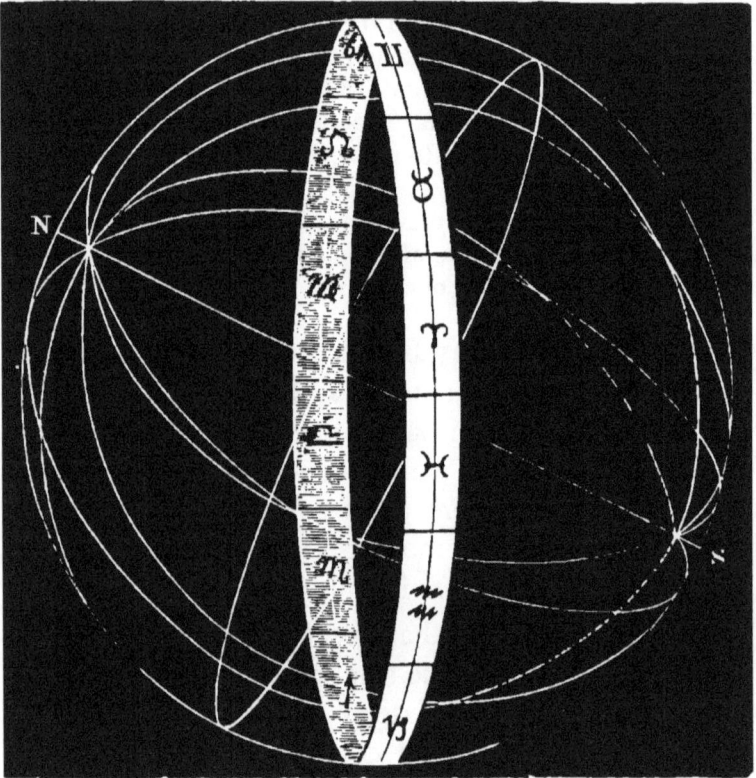

Fig. 20.   Zodiac encircling the Heavens.

zodiac, with the ecliptic running along the middle. Nearly the whole of the sign Cancer (♋) is seen, with a portion of Leo (♌) and a corner of Gemini (♊). (See Appendix IV.)

On the 21st of June we see the sun entering the

sign Cancer. When he sets the next day, he will have advanced nearly one degree *into* Cancer — having moved along the ecliptic to *b*, as shown in § 79 — leaving the star *a* below the horizon. A month later he will have moved along the ecliptic to Leo (♌). At sunset at this time the bright stars seen in the engraving will be below the horizon, and those above, or east of them, will have taken their place. During the next month the sun will enter Virgo (♍), the next sign east of Leo.

†**Solar and Sidereal Day.** — *Questions:* 1. Suppose the sun and a certain star, *a*, Fig. 21, set at the same instant to-day; which will set first to-morrow?

2. What will be the difference in their time of setting? — *Ans.* The sun, being $\frac{1}{365}$ of the whole circle around the heavens *east* of the star, will set $\frac{1}{365}$ of 24 hours, or nearly four minutes, later.

Hence, the time between sunset and sunset is about four minutes longer than that between starset and starset. The former is called a *solar* day, and the latter a *sidereal* day. Each is divided into twenty-four equal parts, called in one case solar hours, and in the other sidereal hours.

3. How many solar hours and minutes are there in a sidereal day? — *Ans.* 23h. 56 min. (nearly).

4. How many sidereal days in 365 solar days? — *Ans.* 366 (nearly).

84. **An Effective Illustration** of this whole subject may be made as follows: Let the blackboards around the schoolroom represent the zodiacal belt around the celestial sphere. Let the head of a pupil standing in

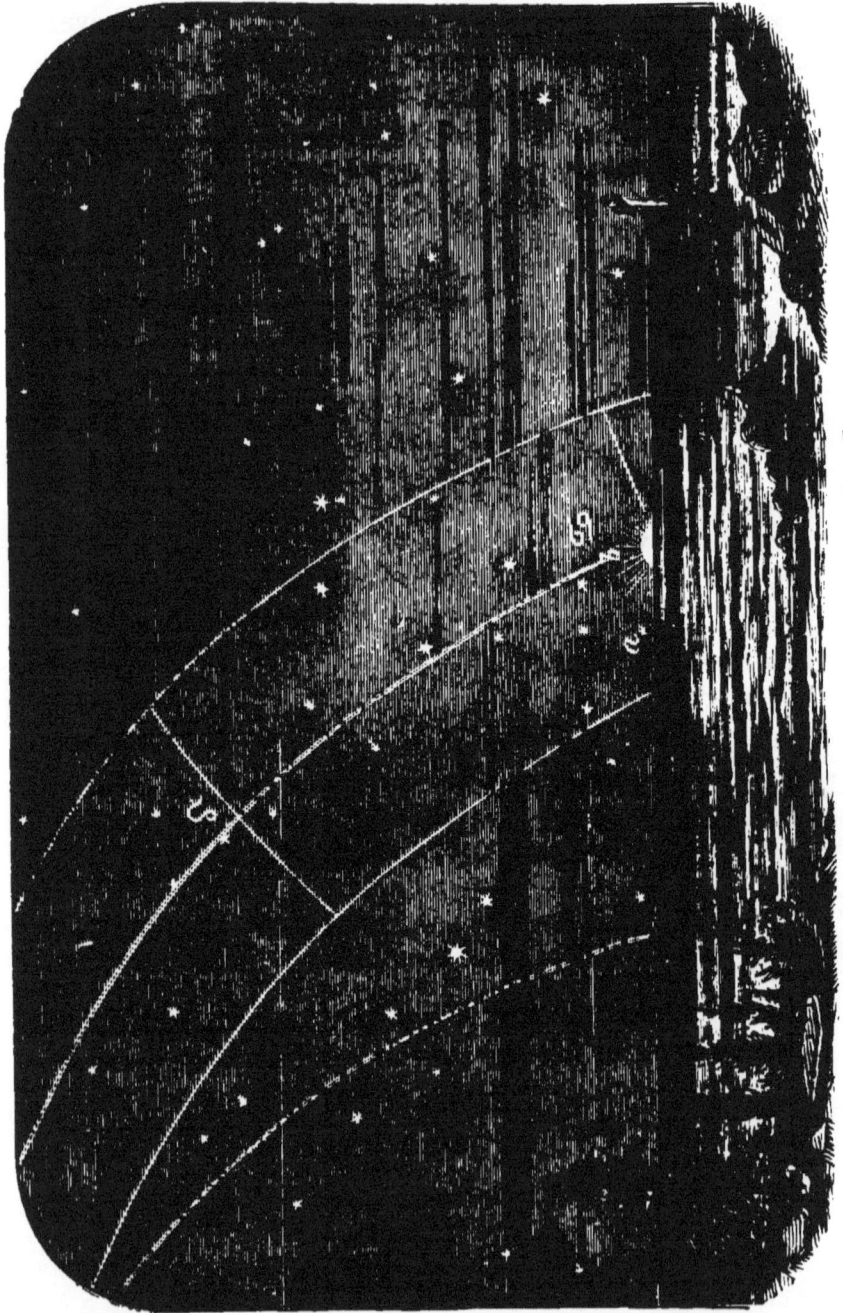

Fig. 21.   Sunset, June 21st.   Sun entering Cancer.

the centre of the room represent the sun. Then let another pupil, carrying a globe with its axis properly inclined, walk around the "sun" to represent the earth moving in its orbit. Let a third pupil walk along by the blackboards and mark the points against which the second pupil sees the "sun" projected from his different positions. A continuous line connecting these points will represent the ecliptic passing along the middle of the zodiac. Then let the zodiacal signs be properly placed with "Cancer" and "Capricornus" over their respective tropics on the globe. Let the globe rotate once on its axis while its bearer advances a step in the orbit to show how the sun seems to pass one degree along the ecliptic in one day. (See Appendix V.)

† **EXERCISES.**

1. What distinguishes the constellations of the zodiac from the other constellations in the sky? (§ 82.)

2. If the sun is exactly on the meridian at this moment, where will it be in exactly twenty-four hours? Where will it be in exactly one sidereal day?

3. As you ride in the cars, what objects seem to move past you more rapidly, those which are near, or those which are more distant? Then, if you had no other means of judging of distance, which would you conclude to be farther off, those which seem to move slowly, or those which seem to move rapidly? What would you conclude to be the distance of hill-tops which seem not to move past you at all?

4. The earth is carrying you along in its orbit as the car carries you along over the rails (having, at the same time, another motion which the car has not; viz., rotation); you see the sun and stars in the zodiac, as you see various objects in the landscape from the car window. The sun seems to change its place as you move; the stars do not. What may you conclude from this alone, in regard to their comparative distances?

IX. THE INCLINATION OF THE EARTH'S AXIS.

85. **Plane of the Earth's Orbit, or Plane of the Ecliptic.**—Suppose two spheres representing the earth and the sun to be half immersed in a smooth sheet of water, the former floating around the latter in an elliptical orbit, shown by the dark line (Fig. 22). The smooth surface of the water represents a plane passing through the earth's orbit. Now, if you can imagine the water removed and the spheres still continuing their motions undisturbed, the exact space which the surface of the water occupied will present to your mind a very accurate idea of the plane of the earth's orbit, or plane of the ecliptic. It must be imagined as cutting through the centres of both sun and earth, and extending to an indefinite distance beyond the earth's orbit. To an observer standing upon the ball representing the earth the other ball would seem to move around him in the surface of the water ; hence the plane of the earth's orbit is also the plane of the sun's apparent path, or ecliptic. (Appendix VI.)

86. **The Earth's Axis is inclined to the Plane of the Ecliptic.**— We observe that the earth's axis, as represented in Fig. 22, does not stand upright in the plane of the ecliptic, and that the equator is, therefore, cut by the plane in two points.

The Amount of the Inclination of the Earth's Axis to a perpendicular to the plane of the ecliptic is 23½ degrees, or 66½ degrees to the plane itself.

Fig. 2. Illustration of the Plane of the Ecliptic.

Let NESW represent the earth, and the line S'S' the plane of the ecliptic with its edge turned exactly towards us.   If the axis were perpendicular to the ecliptic, it would be represented by the line PR, and TT' would represent the equator lying exactly in the plane, or, as the usual expression is, " coinciding with

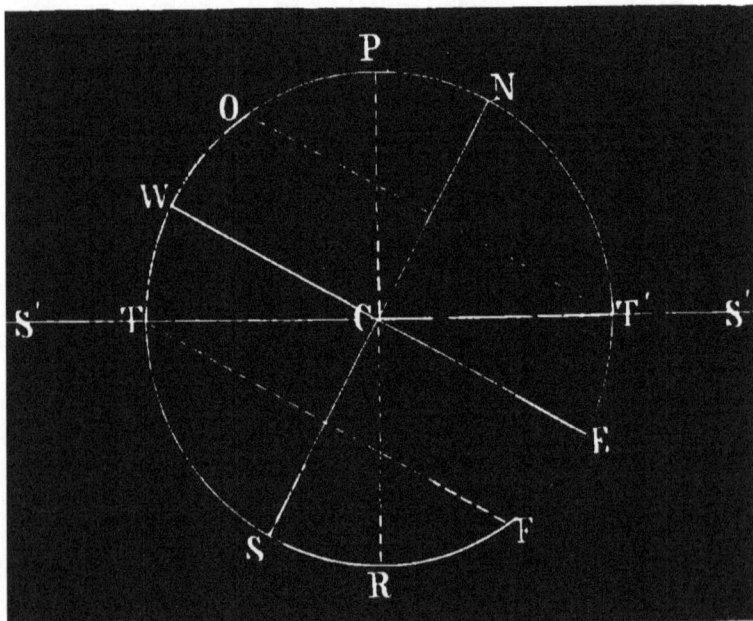

Fix. 23. Inclination of the Earth's Axis.

it."   If, now, we take the north pole, P, and move it $23\frac{1}{2}$ degrees to the point N, we shall have it in its true position.   While we are doing this we move every other point in the whole circumference an equal distance ; the point R moves to S, and the points TT' of the equator move to W and E, each $23\frac{1}{2}$ degrees from the plane of the ecliptic.

87. *Relation of the Tropics to the Equator and to the Plane of the Ecliptic.* — The tropics of Cancer and Capricorn are denoted by the circles on either side of the equator in Fig. 22, and by the lines OT′ and TF in Fig. 23. They are each 23½ degrees from the equator, and we see that the plane of the ecliptic cuts across from one to the other.

A circle called the "Ecliptic" is generally drawn upon terrestrial globes across the equator from tropic to tropic. A similar circle is made by the surface of the water around the globe represented in Fig. 22, which explains the meaning of the former. The circle drawn upon the globe must rotate with the globe, however, whereas it should be stationary, like the circle made by the surface of the water.

88. **The Earth's Axis constantly points in the Same Direction during the Yearly Revolution; viz., towards the North Star.** — If it were not for this fact we should not always see the north star, summer and winter, at the same distance above the northern horizon.

This unchanging direction of the earth's axis is exhibited in Fig. 26. The north star must be imagined at an immense distance in the direction in which the various lines denoting the earth's axis point; so that, although the lines, being parallel, are really directed to four different points, yet, like the parallel lines of a railway track, they seem to meet in a single point in the distance.

89. **How we know that the Earth's Axis is inclined to the Plane of the Ecliptic.** — The fact of the

inclination is proved, and its amount measured, by the sun's apparent movements north and south of the equator, treated of in the following section.

THE SUN'S DECLINATIONS, OR APPARENT MOVEMENTS NORTH AND SOUTH OF THE EQUATOR.

90. **Cause of the Sun's Declinations.** — Every one is familiar with the expressions, " The sun is crossing the line " ; " The sun is coming north, and we shall soon have warm weather " ; " The sun is going south, and the days are growing shorter " ; etc. Like the two apparent movements of the sun already described, this is not due to any change in the sun's real position, but must come home to the earth itself. The cause is threefold : —

(1)   The inclination of the earth's axis.

(2)   The unchanging direction of the axis.

(3)   The earth's revolution around the sun.

As a result of these three conditions, the north pole of the earth is sometimes inclined directly *towards* the sun, at which time the sun is over the tropic of Cancer, $23\frac{1}{2}$ degrees north of the equator. At other times the north pole is inclined directly *from* the sun, at which time the sun is over the tropic of Capricorn, $23\frac{1}{2}$ degrees south of the equator. During the intermediate times the sun must be somewhere between these circles, being directly over the equator, or " crossing the line," twice a year.

91. *The Sun's Declinations appear in a Spiral Path winding around the Sky, like the Threads of a Screw.* — This is in consequence of the earth's two motions

going on together. On the 20th of March we see the
sun rise at E, Fig. 24, describe the arc through A, and
set at C. This arc is directly over the equator, and is,
therefore, the *equinoctial.* Day and night are now of
equal length, and the period is for this reason styled

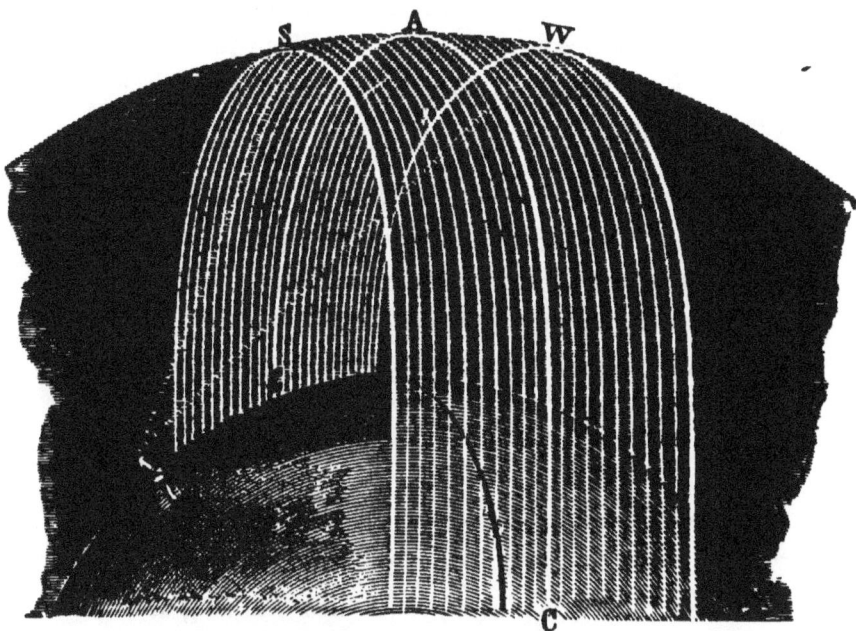

Fig. 24. Sun's Apparent Motion on Different Days during
the Year, North and South of the Equator.

the *spring,* or *vernal, equinox.* On the next day the
sun describes a circle a little north of the equinoctial;
the next, still farther north; and so on, until on the
21st of June he has reached the limit of his northern
declination. The circle which he describes on this
day, S, is directly over the tropic of Cancer, $23\frac{1}{2}$ degrees north of the equator, or equinoctial. For a few

days he seems to describe nearly the same circle, whence the name of the period, *summer solstice.**

From the summer solstice the sun describes his circles farther and farther south each day, until on the 23d of September he again describes the equinoctial, EAC.   Day and night are again equal—the *autumnal equinox.*   Thence he continues still southward till the 21st of December, when he describes the circle W over the tropic of Capricorn.   This is the limit of his southern declination, the *winter solstice,* from which after a few days he begins his return northward.

In Fig. 21 the oblique line at the left of the zodiac represents a portion of the equinoctial.   The sun is seen at his greatest distance north (the summer solstice), and as he moves along the ecliptic toward ♌ day after day, it is plain that he will constantly approach the equinoctial, until he will cross it at the autumnal equinox.

92. *Tropics named from the Signs Cancer and Capricornus.* — On the 21st of June, when the sun is over the tropic of Cancer, he is also entering the sign Cancer; on the 21st of December, when he is over the tropic of Capricorn, he is entering the sign Capricornus.

93. **The Effects of the Sun's Northern and Southern Declinations:—**

(1) The Change of Seasons.

(2) The Variation in the Length of Day and Night.

---

* *Solstice* means *stationary sun,*

THE CHANGE OF SEASONS. — THE VARIATION IN THE
LENGTH OF DAY AND NIGHT.

94. **Suppose the Earth's Axis were Perpendicular
to the Plane of the Ecliptic** (Fig. 25), then the plane
of the ecliptic would coincide with the equator, and
the sun would, accordingly, always be seen over the
equator. His northernmost rays would always strike
exactly at the north pole ; his southernmost rays, at
the south pole. Those rays which reach us would

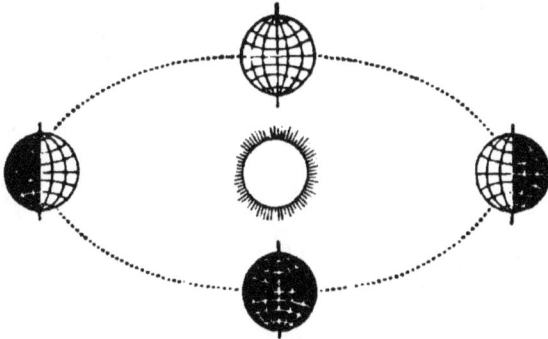

Fig. 25. Axis Perpendicular to Ecliptic.

come in precisely the same direction every day throughout the year, and there would be nothing to produce a
change of seasons except the difference in our distance
from the sun at different points of the orbit (§ 41).
This difference, however, is so extremely slight that in
all probability ordinary observers could not detect the
consequent difference of temperature,* and conse-

---

* Whatever difference there might be would be just opposite to
that produced by the sun's declinations, at least in the northern
hemisphere, for we are nearest the sun in winter.

quently there would be perpetual winter in the frigid zones, perpetual spring in the temperate zones, and perpetual summer at the equator.

Every parallel circle would be half the time in the sunlight and half the time in the shade, so that day and night would be equal throughout the year at every

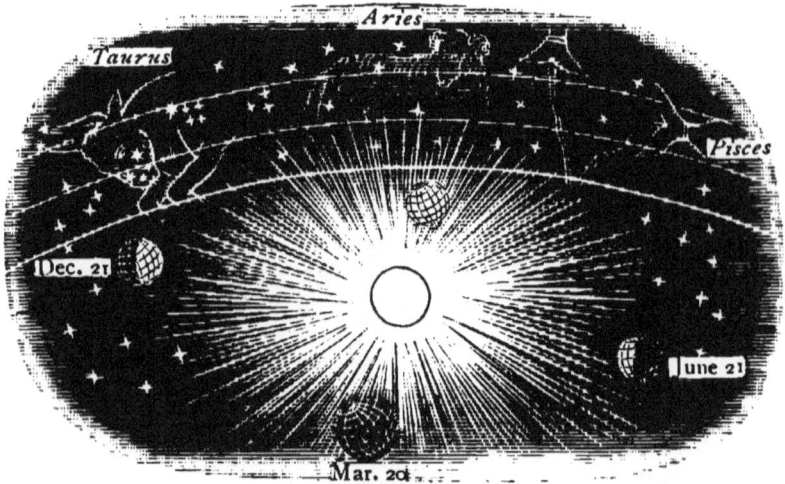

Fig. 26.   The Change of Seasons.

point on the earth's surface — a perpetual equinox — except at the poles, where the sun would always be seen in the horizon.

95. **The Change of Seasons.**—We have seen (§ 54) that the change of seasons is due to differences in the direction of the sun's rays, which beat almost directly upon our heads in midsummer, and fall very obliquely in midwinter Also (§ 90) that these differences of direction, in other words the sun's declinations, are produced by the inclination and unchanging direction

of the earth's axis, together with the earth's yearly revolution around the sun.

Any two of these three conditions may be conceived to exist without necessarily producing a change of seasons. For example, we may imagine the earth's axis as always preserving its present inclination of 23½ degrees, while at the same time its north pole inclines exactly towards the sun during the entire yearly revolution. In this case the sun would, of course, always be vertical at the tropic of Cancer, causing a perpetual summer in the northern hemisphere, and a perpetual winter in the southern hemisphere.

96. **The Variation in the Length of Day and Night.** — This change is due to the same cause that produces the change of seasons ; for, when the sun is north or south of the equator, more or less than half of each parallel, except the equator, is in the sunlight at a time, as may be seen from Figs. 28 and 30, or from the right and left of Fig. 26. At these times, therefore, we are more or less than twelve hours in passing through the sunlight or shade.

(The same truth was shown in a different manner by Fig. 17.)

Both summer heat and winter cold are thus increased ; for the sun shines upon us in summer not only more directly, but for more hours at a time than in winter.

Let us now examine, in regular order, the changing relations which the earth bears to the sun as it moves in its orbit.

97. **The Vernal Equinox,** *20th of March.* — The sun enters the sign Aries. (See bottom of Fig. 26. The sun as seen from the earth in this part of its orbit appears in Aries.) Neither pole of the earth inclines towards or from the sun, but *sidewise ;* hence the sun is vertical at the equator, his northernmost and south-

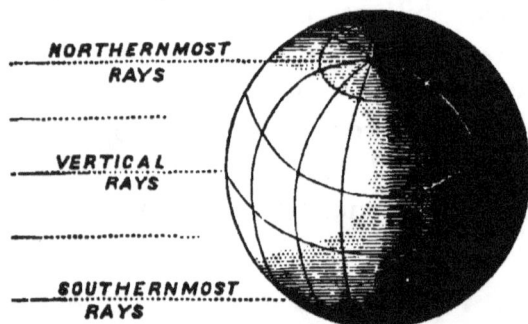

NORTHERNMOST
RAYS

VERTICAL
RAYS

SOUTHERNMOST
RAYS

Fig. 27.   Earth at Vernal Equinox.

ernmost rays fall at the poles, and day and night are everywhere equal. It is spring in the northern and autumn in the southern hemisphere.

As the earth moves on from the vernal equinox, the north pole begins to lean towards the sun and the south pole from it ; the sun, therefore, is vertical farther and farther north of the equator each day (see Fig. 26), until —

98. **The Summer Solstice,** *21st of June.* — The sun now enters Cancer. The north pole leans exactly towards the sun, and the south pole exactly from it ; consequently, the sun is vertical $23\frac{1}{2}$ degrees north of the equator, and, if his vertical rays should leave a track as the earth turns upon its axis, they would mark the tropic of Cancer upon the surface (Fig. 28).

The northernmost ray of the sun falls 23½ degrees beyond the north pole, and would if it left a track mark the arctic circle as the earth turns upon its axis. The southernmost ray falls 23½ degrees short of the south pole, and describes the antarctic circle.

No point within the arctic circle passes out of the sunlight, and no point within the antarctic circle passes into the sunlight, during the earth's rotation ; hence a 24-hours' day within the former, and a 24-hours' night

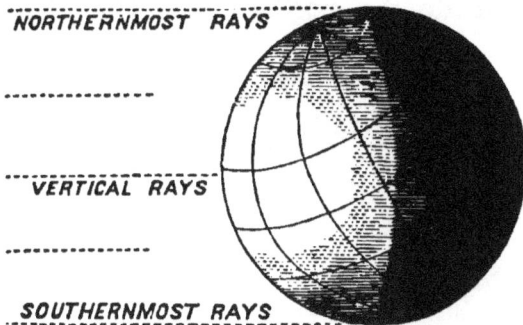

Fig. 28. Earth at Summer Solstice.

within the latter. The long *twilight*, however, practically shortens the night very much in the antarctic circle, excepting in a small space around the pole. (See p. 25, note at bottom.)

More than half of each parallel circle in the northern hemisphere is in the sunlight at a time, and less than half of each parallel circle in the southern hemisphere ; hence the days are longer than the nights in the former, and shorter than the nights in the latter. It is summer in the northern and winter in the southern hemisphere.

From the summer solstice, the poles incline less and less towards and from the sun, and the sun's vertical rays fall farther and farther south, until —

99. **The Autumnal Equinox,** *23d of September.* — QUESTIONS : —

1. What season is it in the northern hemisphere? In the southern?

2. How do the poles lean with reference to the sun? (Fig. 29.)

3. Where do the northernmost and southernmost

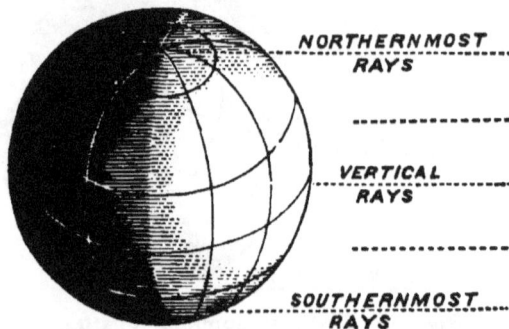

Fig. 29.   **Earth at Autumnal Equinox.**

rays of the sun fall? Then on what circle of the earth do the vertical rays fall?

4. What is the length of day and night?

100. **The Winter Solstice,** *21st of December.* — QUESTIONS : —

1. What season in each hemisphere?

2. At what circle of the earth is the sun vertical?

3. Which pole leans exactly towards the sun?

4. Where do the northernmost and southernmost rays fall?

5. In which zone is there a 24-hours' day? In which a 24-hours' night?

6. How does the day compare with the night in each hemisphere?

ADDITIONAL OBSERVATIONS.

101. **Day and Night at the Poles** are each six months in length. From March 20th to September

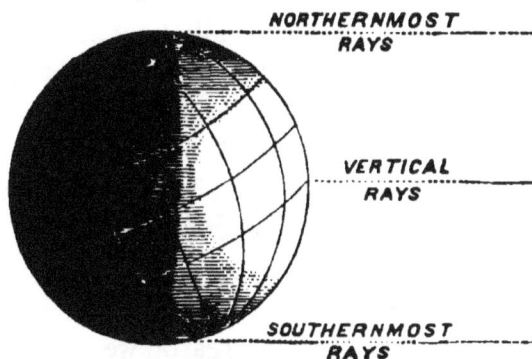

Fig. 30. Earth at Winter Solstice.

23d the north pole remains constantly in the sunlight, and the south pole in the shade. From September 23d to March 20th the conditions are reversed.

We arrive at the same result by reflecting that the poles are not affected by the earth's daily motion, and that if it were not for this motion the day and night would be six months each throughout the year over the whole earth.

Within the frigid zones, day and night each vary all the way from six months at the poles to twenty-four hours at the polar circles.

102. **Seasons at the Equator.** — The sun crosses the equator and departs to its greatest distance from the equator twice during the year. There are, therefore, two summers and two winters annually at the equator, although "winter" there must, of course, be much warmer than our warmest summer.*

103. **The Full Effects of the Various Changes in the Direction of the Sun's Rays are not felt at once.** — Although the most direct rays fall at noon, the warmest part of the day is usually two or three hours later. So, although the hottest rays fall at the summer solstice yet our warmest weather does not come until some time afterwards. We continue to receive more heat during the days following than we lose during the nights. Thus the great heat of a July or August day is not produced entirely by the sun of that day, but is an accumulation of the heat of the several preceding weeks. For a like reason we do not experience the greatest cold at the winter solstice. We continue to lose more heat during the night than we receive during the day, and the maximum of cold does not arrive until some time in January.

EXERCISES.

1. Would there be any tropics or polar circles if the earth's axis were not inclined from a perpendicular to the ecliptic?
2. Where would the tropics be if the axis were inclined 45°? Where would the polar circles be?

---

* The only seasons practically known in tropical climates are the "wet" and the "dry."

3. How would the extremes of heat and cold compare with those we experience?

4. How much should the axis be inclined to bring the tropic of Cancer to New York?

5. How much should the axis be inclined to bring the arctic circle to New York?

6. If the axis were thus inclined, how long would be the day and the night at the summer solstice at New York? Where would the sun be seen at the winter solstice?

EXERCISES FOR REVIEW.

1. Of what is the earth a part?
2. How large a part?
3. Prove that it is spherical.
4. What made it so?
5. Prove that it is spheroidal.
6. What made it so?
7. What is its diameter?
8. Its circumference?
9. How many square miles in its surface?
10. How much is it flattened at its poles?
11. What fixes the position of its equator and poles?
12. How many different points upon its surface are in the same latitude?
13. In the same longitude?
14. In the same latitude *and* longitude?
15. What is the prime meridian?
16. What may be called the *prime parallel?*
17. Which is the longer, a degree of the parallel passing through London, or of that passing through Washington?
18. What zones are not *belts?*
19. Give the breadth of each zone in degrees.
20. What circles separate the zones, and what fixes the position of these circles?
21. How would the sky appear by day if it were not for the air?
22. Why is it not dark the moment the sun has set?
23. Why is it not as light and warm at sunrise as at noon?
24. Describe the earth's motions.
25. Why do they not cease?
26. Prove that it rotates.
27. Effects of its rotation.
28. Which *seems* to rotate, the earth or the starry sphere?

29. Would this be the appearance to a spectator in space?
30. What prevents the earth from flying off into space?
31. From falling to the sun?
32. What is the form of its orbit?
33. Define *perihelion* and *aphelion.*
34. What are the effects of its yearly motion?
35. If the sun were visible at the same time with the stars, would it always appear in the same place among them?
36. How much would it appear to move in one day?
37. In what direction?
38. Along what line?
39. Through what belt of the starry heavens?
40. What is the cause of this apparent motion?
41. Why is a picture of a ram placed upon the Almanac page for March?
42. What is the name of the plane cutting through the centre of the earth and through the ecliptic?
43. Is the earth's axis perpendicular to this plane?
44. What would be the results if it were so?
45. What is its true attitude in the plane?
46. When does the north pole lean exactly towards the sun?
47. Upon which circle of the earth do his vertical rays then fall?
48. Is the sun then high or low in the heavens to us at noon?
49. How does day compare with night in length?
50. What, then, is the season at this time?
51. When does the north pole lean exactly from the sun? (*Repeat Questions* 47, 48, 49, 50.)
52. When do the poles lean neither towards nor from the sun? (*Repeat Questions* 47, 48, 49, 50.)
53. What is the length of day and night at the poles?
54. Of the longest day and night at the polar circles?
55. At the equator?

APPENDIX.

I. (§ 40, p. 16). **The Exact Polar Diameter of the Earth** is 7,899.58 miles, while the equatorial diameter is 26.48 miles greater, or 7,926.59 miles.

Multiplying these numbers by 3.1416, we have the circumference of a meridian circle, or polar circumference, 24,817 miles and the equatorial circumference, 24,902 miles.

II. (§ 42, p. 16). **The Moon's Distance from the Earth** varies during the month, the greatest distance (*Apogee*) being about 252,000 miles; and the least (*Perigee*), about 226,000 miles. Its mean distance is, therefore, about 239,000 miles.

For the sake of comparing the earth's magnitude with that of other heavenly bodies, and with space, other distances and magnitudes are annexed.

The Distances of the Other Known Planets from the sun vary from about one-third to thirty times that of the earth.

Ex. — About how many miles from the sun is the nearest planet? The most distant planet?

The Distances of the Fixed Stars are so great that they are utterly beyond our comprehension. The very nearest of them is about 200,000 times more distant from us than we are from the sun. But even this vast distance is small compared with that of the great multitude of the stars.

The Sun's Diameter is 860,000 miles, or nearly 108 times as great as that of the earth. Therefore, as spheres are to one another as the cubes of their diameters, the sun must be a body more than 1,250,000 ($108 \times 108 \times 108$) times as large as the earth.

`The Moon's Diameter** is 2,162.5 miles, or a little more than one-fourth that of the earth.

The Diameters of the Other Known Planets vary from a little less than one-half to more than eleven times that of the earth.

Densities of the Other Heavenly Bodies. — The sun's density is 1½; that of the moon is 3½; densities of the other planets vary from 1 to 7.

Density of the Earth. — The earth weighs about 5½ times as much as it would weigh if composed entirely of water. We say, therefore, that its density is 5½.

III. (§ 65, p. 31).   † **Foucault's Experiment proving the Earth's Rotation.** — Attach a pendulum to a large globe so that the point of suspension shall be over its pole; let the pendulum end in a sharp point which will make a scratch upon the globe at each vibration; let the pendulum swing, and slowly rotate the globe under it. You will observe that, notwithstanding the rotation, the pendulum will constantly swing towards the same two points in the room; that is, in the same plane. The consequence will be a star-shaped figure

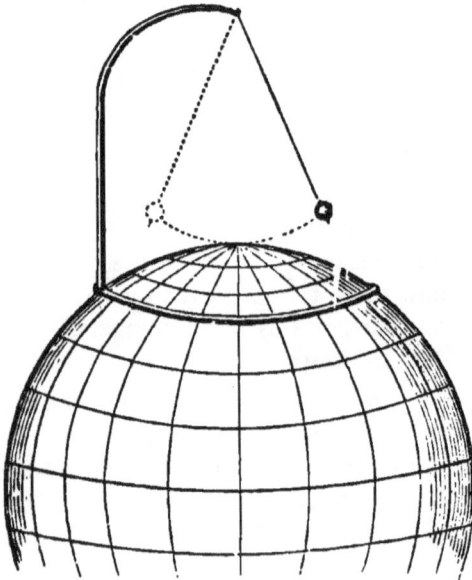

Fig. 31.   Proof of the Earth's Rotation.

scratched upon the globe by the pendulum point, which will make a different line at each vibration. A similar experiment has been tried upon the earth itself, with a like result. At the equator, where the relation between the plane of vibration and the earth's surface is not changed by the rotation, the pendulum marks only one line; but the nearer it is to the pole, the nearer the figure which it describes approaches the star-shaped figure represented in the engraving, which could not be the case if the earth did not rotate.

IV. (§ 82, p. 45). † The Signs of the Zodiac, with their Symbols, and the months in which the sun appears in each respectively.

1.	Aries,	*the Ram,*	♈	*March.*
2.	Taurus,	*the Bull,*	♉	*April.*
3.	Gemini,	*the Twins,*	♊	*May.*
4.	Cancer,	*the Crab,*	♋	*June.*
5.	Leo,	*the Lion,*	♌	*July.*
6.	Virgo,	*the Virgin,*	♍	*August.*
7.	Libra,	*the Balance,*	♎	*September.*
8.	Scorpio,	*the Scorpion,*	♏	*October.*
9.	Sagittarius,	*the Archer,*	♐	*November.*
10.	Capricornus,	*the Goat,*	♑	*December.*
11.	Aquarius,	*the Waterbearer,*	♒	*January.*
12.	Pisces,	*the Fishes,*	♓	*February.*

V. (§ 84, p. 49). **A Field Experiment.** —Select a tree in a broad, open field to represent the sun, and station yourself at a little distance from it to represent the earth. Now, as the stars are inconceivably more distant than the sun, they must be represented by comparatively distant objects, as, for example, those in the horizon. Suppose, therefore, the tree-tops, church-spires, hills, etc., in the horizon to be stars in the zodiac surrounding the sun and earth, like the circle in Fig. 19, p. 44.

Now in the first place imitate the daily motion of the earth *alone* by turning slowly on your heels without moving from your place. All objects in sight seem to revolve around you in the direction opposite to that in which you are turning, and not only this, but all seem to perform their revolutions in *the same time;* the objects in the horizon seem to describe their *great* circles as quickly as the tree describes its *small* circle. More-over, you always see the tree against the same point in the hori-zon. From this you infer that, if the earth only rotated on its axis without moving from its place, the sun and stars would per-form their apparent daily revolutions around the earth in pre-cisely the same time, and that the sun would always be seen among the same stars.

Now imitate the yearly motion taking place alone, by moving in a circle around the tree without turning on your heels; that is, always facing in the same direction, as, for example, south or east.

Ah! now a great difference is seen. The tree seems to re-volve around you in the same direction in which you revolve around the tree, only at the opposite point in the circle, and if you watch it against the horizon you will see it moving past the tree-tops, church-spires, etc., completing its apparent revolution in the same time that you complete your real revolution. Ob-serve also that it appears on one side of you, passes in front, and disappears on the other side, while you are performing *half* your revolution, and remains out of sight during the other half. These appearances teach you that if the earth performed its revolution around the sun without rotating on its axis, the sun would rise, perform a six months' journey through the constella-tions, and then disappear for the remaining six months of the year. You notice also that the tree-tops, etc., in the horizon do not appear to change their positions in the least perceptible degree during your revolution around the tree, but that you see them in the same direction from all sides of your orbit. This illustrates to you the significance of the word *fixed* as applied to the stars, which, were it not for the daily rotation of the earth, would always remain fixed in the same points of the sky, as far as ordinary vision could determine. A nice instrument would enable you to distinguish a slight *parallax*, or change of position, in some of the objects in the horizon as viewed from opposite points of your orbit around the tree; but a much more delicate instrument would be required to detect the parallax of the stars as seen from opposite points of the earth's orbit.

Having imitated the daily and yearly motions separately, now imitate them together, as the earth performs them. Of course, the two classes of *effects* will be combined, and will correspond exactly with those which you observe in the heavens. Every time you turn round to the tree (*sun*) you find it has made a little advance in the horizon (*ecliptic*), until it has described the whole circle.

VI. (§ 85, p. 50). † **Meaning of "Ecliptic."** — The moon revolves around the earth in an orbit which crosses the plane of the ecliptic at a small angle, so that it is half the time on one side, and the other half on the other side of this plane. The smallest ball in the engraving (Fig. 22) represents the moon thus revolving around the earth and crossing the plane of the ecliptic in two points (*nodes*). Now, no *eclipse*, either of the sun or moon, can take place excepting when the moon is crossing the plane, as is evident from the figure. Hence, the plane of the earth's orbit takes its other name — *plane of the ecliptic* (or *eclipses*).

The Astronomical Lantern.

The object of the Astronomical Lantern is to facilitate the study of stellar astronomy. It is in tended for beginners, for astronomical classes in high schools or private schools, and, in fact, for all who desire to become acquainted with the constel lations.

The difficulty hitherto experienced in this study, and which is obviated by the use of the Lantern, is this : In order to study the starry heavens, it has been necessary to use an astronomical atlas, or a celestial globe. These must be examined in the house, by the light of a lamp. The observer, having found his constellation on the atlas, goes out to look for it in the sky. But, by the time he gets out of doors, he has forgotten how it looked on the atlas. And when he has found it in the sky, he has forgotten how it looked there, before he gets back to his atlas or globe. All who have studied the constellations have met with this difficulty.

Now, the Astronomical Lantern makes the study of the stars perfectly simple and easy. It is constructed like a dark-lantern, closed on three sides, and on the fourth provided with a ground glass, in front of which maps can be inserted. On each of these maps, which are semi-transparent, is represented a constella'ion, the places of the stars being indicated by perforations, through which the light shines. The largest perforations are for the stars of the first magnitude, and the smaller, in due proportion, for the lesser stars. The student, therefore, wishing to observe any particular constellation or cluster, has only to light a candle within the Lantern, insert the appropriate slide, and go out into the night. Holding up the Lantern in one hand, he can compare the constellation as it appears on the Lantern with that in the sky, until he becomes perfectly familiar with the latter.

It is easy to see how much the use of such a Lantern facilitates the whole study. In fact, we think that henceforth no one wishing to become acquainted with the heavens can afford to dispense with it. The increased ease of the study will probably also enlarge the number of students in this interesting department of science.

To use the Lantern, it is necessary to see what constellations are favorably situated for observation at the time; which can be done by the help of Dr. Clarke's manual, "How to Find the Stars," which accompanies every Lantern sold.

The card-slides accompanying the Lantern are seventeen in number, and contain all the constellations visible to an observer in the North Temperate Zone. Other slides may easily be added as required. In these maps of the constellations the names and the designations of the stars are retained, but the figures of bears, bulls, unicorns, sheep, virgins, dragons, lions, and the like, which have so long disfigured the celestial globe, are omitted. Instead of these confusing figures, few of which bear any resemblance to the constellations, we have substituted dotted lines, tying together in simple diagrams the chief stars in each cluster. Experience shows that by these diagrams the separate constellations are much more easily recognized and remembered than by the traditional pictures of animals, monsters, and men, which have hitherto crowded the starry atlas. By these connecting lines, too, the principal stars in each group are easily found and associated in the memory.

In preparing these maps, we have followed the "Uranometria Nova" of Argelander. This atlas was selected because of its reputation for accuracy, and because the scale by which it is drawn was best adapted to the size of the slides. At the top of the map are given the names of the constellations which it contains. At the bottom is given their position in the heavens at such time of the year as is suitable for observation. The stars are lettered with their proper symbol. Double stars are indicated by a D. The nebulæ are shown by means of a group of minute dots, and star-clusters in a similar way. On each map there is also a list of the telescopic objects which are to be found in the constellations represented upon it, — those, at least, which are suitable for small

telescopes. In this way the Lantern may be of great
use to observers possessing such instruments, by
enabling them to find easily the double stars, clusters, etc., which are in a convenient position for
observation at any period of the year. Those who
have spent hours in looking through books of astronomy, in order to see what suitable subjects for their
telescopes are above the horizon at any particular
time, will easily understand the advantage of this
arrangement. — *From* DR. CLARKE's "*How to Find
the Stars.*"

HOW TO FIND THE STARS.

The object of this little book is to help the beginner to
become better acquainted, in the easiest way, with the visible
starry heavens; to know the winter and summer constellations, and the principal fixed stars. It shows the position of
the constellations at different periods of the year, giving their
place in each of the four seasons. It also shows how to find
the separate clusters by a series of triangles and diagrams,
covering the whole heavens, and connecting each constellation with its neighbors. It indicates the most interesting
objects at each period of the year, especially such as can be
found with a telescope of moderate power. It closes with a
description of the Astronomical Lantern.

*The former price of the Lantern was $6.00; we now offer
it, in improved form, with seventeen slides and a copy of
"*How to Find the Stars,*" for $4.50. The latter is also
sold separately at 15 cents per copy.*

D. C. HEATH & CO., Publishers,
5 SOMERSET STREET, BOSTON.

5

The following testimonials from those who have used this Lantern have been recently received: —

C. A. Young, *Prof. of Astronomy, Princeton College:* I have carefully examined Dr. Freeman Clarke's Astronomical Lantern, and find it to be an admirably contrived apparatus for its purpose,—simple, easily managed, and effective. I think an adequate knowledge of the constellations could be obtained by its use, in connection with the little book that accompanies it, more rapidly and easily than from the most elaborate and expensive celestial globe. (*Aug.* 8, 1885.)

C. S. Lyman, *Prof. of Astronomy, Yale College:* Dr. Clarke's Lantern is certainly a very simple and happy device for facilitating the study of the stellar configurations, and aiding the student in familiarizing himself with the nightly aspects of the heavens. The mere study of the constellations is, indeed, but a small part of astronomy, yet a part both interesting and important to the beginner; and I have never known any contrivance that could compare with this Lantern for saving alike time, trouble, and eye-sight, and rendering such study attractive and easy. It is such a device as the writer well remembers often desiring, and even purposing to construct, yet never brought to pass. Schools, academies, colleges, and amateur astronomers cannot fail to find it useful, and all who use it will feel thankful to Dr. Clarke and his publishers for putting so convenient a piece of apparatus within their reach. (*Dec.* 15, 1885.)

O. C. Wendell, *Harvard Coll. Observatory:* I have examined the Astronomical Lantern, and find it well adapted to its purpose. It combines in a high degree simplicity with clearness, and, for beginners and amateurs, I think it has no equal. Among its salient features are its giving bright stars on a dark field, together with the fact that it represents the magnitudes of the stars by the size of the apertures. This is a great help to young people, from its naturalness. Besides, the diffused light transmitted through the cardboard enables one easily to see the printed magnitudes, as well as letters and constellations. Mr. Clarke seems to have done a real service to the youthful patrons of astronomy in devising a lantern which is at once portable, highly entertaining, and so cheap as to be within the reach of all. I am confident that the names of the principal stars and constellations can be learned from it in half the time required to learn them from an ordinary map. (*Nov.* 18, 1885.)

C. Getchell, *Science Teacher, Phillips Exeter Acad.:* I consider Dr. Clarke's Astronomical Lantern the best means for
enabling students to identify in the heavens those stars which
they have studied from a map or globe. In the recitation-room
it shows to the whole class the constellations in the same positions in which they appear in the sky. In case the teacher
does not have the opportunity for much out-door work, good
results are obtained by requiring each pupil to make for himself
a set of perforated maps from his own observation. These
maps may be verified by comparison with those furnished with
the Lantern. (*June* 29, 1885.)

E. H. Rudd, *Science Teacher, St. Mary's School, Knoxville,
Ill.:* I have used Dr. Clarke's Astronomical Lantern for
two years, and find it of the greatest practical benefit to my
classes in astronomy. There is nothing that I know of which
could take its place. The interesting part of astronomy to most
pupils is learning the names, form, and location of constellations. My experience is, that the Lantern, for this purpose, is
worth a dozen globes or charts. Nothing, to my mind, is better
than the system of triangulation as a help to location.
(*Aug.* 1, 1885.)

R. W. B. Elliott, *San Antonio, Tex.:* I have found the
Astronomical Lantern, invented by the Rev. Dr. J. F. Clarke,
very useful in arousing the interest of my children, and enabling
them to identify the different stars and constellations. It furnishes a very pleasant and instructive means of passing the
earlier parts of our bright Texas nights. (*July* 27, 1885.)

Mary A. Brackett, *Prin. Private School, Brooklyn:* I think
it by far the best thing of the kind I have ever seen. Any beginner in the study of practical astronomy could not fail to
derive great pleasure and help from it. Its simplicity gives it
its great charm. (*Aug.* 8, 1885.)

Clement B. Smyth, *New York:* Dr. Clarke's Astronomical
Lantern, one of which I purchased for my daughter, has been of
great interest and aid to her in her studies. It is an apparatus
which, if known to those interested in this study, would, I
should think, be very frequently called for. My daughter is
delighted with it, and feels greatly indebted to her teachers, the
Misses Brackett, of Brooklyn, who kindly advised her of its
value, and where to procure one. (*July* 27, 1885.)

The Stellar Tellurian.

THIS instrument illustrates all the essential principles of Celestial Me-
chanics in so simple and clear a manner that the pupil can understand in a few
hours what he could not grasp in months of mere book study, if at all. It is
accompanied by a **Manual of Direction** describing nearly a
hundred different illustrations, an examination of which cannot but convince
the teacher of the wonderful capabilities of the instrument.

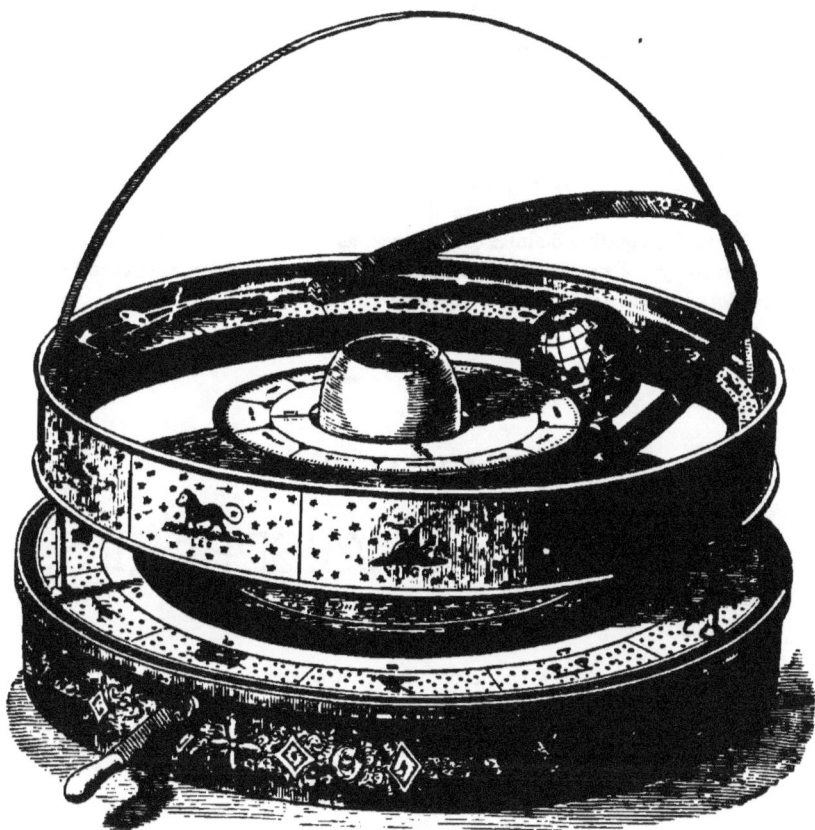

The following are a few of many similar

TESTIMONIALS.

I am satisfied that the Stellar Tellurian is the best instrument of its kind
ever offered to the public. Almost every idea of Astronomy is made so plain
that a child can easily understand it. It should be in all schools. — President Craven, *late of Trinity College, Hartford, Conn.*

(1)

I am well acquainted with the Stellar Tellurian, and consider it a great help in teaching the elements of Astronomy. I think it the best instrument yet made for illustrating the motions of the earth, and the phenomena caused by these motions. I intend to use the instrument at the college as soon as we have more room for apparatus. — J. A. Gillet, *Professor of Physics, New York Normal College.*

I have examined the Stellar Tellurian, and am much pleased with it. I believe in appealing to the eye in giving instruction, where it can be done without giving wrong impressions. I think this is made on correct principles. It is also worthy of note that it appears to be very durable, and, with ordinary care, is not liable to get out of repair. — De Volson Wood, *Professor of Civil Engineering, Stevens Institute (N.J.) of Technology.*

I have been examining to-day the Stellar Tellurian, and am glad to express my opinion of its value. The number of phenomena in Astronomy, which it beautifully illustrates, is very great, and the illustration in each case is as nearly perfect as can be. The use of the Tellurian in the schoolroom must aid the pupil in getting definite conceptions, as well as awaken new interest in Astronomy. *I recommend its use with entire confidence.* — Kendall Brooks, *President Kalamazoo College.*

JACKSON'S

CELESTIAL HEMISPHERES.

THESE are two Wall Maps, each 5 feet in diameter, upon which the pupil may trace the constellations as if he were pointing them out in the sky. With them is a **Key** in which are the Mythological figures, names of constellations, principal stars, etc. Arranged for the Standard Epoch, 1880.

By EDWARD P. JACKSON, A.M. Author of " The Earth in Space : A Manual of Astronomical Geography."

D. C. HEATH & CO., Boston, Mass.

(2)

GUIDES FOR SCIENCE-TEACHING.

INTENDED FOR TEACHERS WHO DESIRE TO PRACTICALLY
INSTRUCT CLASSES IN NATURAL HISTORY.

IN PREPARATION:

D. C. HEATH & CO., Publishers,
BOSTON.